亚太室内设计名师系列丛书　王启贤

PREMIER QUALITY LIVING

优 享 之 旅 · 豪 宅 会 所
WIA DESIGN CONSULTANTS - CLYDE WONG

中国林业出版社　　　　　　　　　　　　　　　　　　　　　创福美图　佳图文化　主编

图书在版编目（CIP）数据

优享之旅 / 佳图文化 主编． -- 北京：中国林业出版社，2016.3

ISBN 978-7-5038-8394-1

Ⅰ．①优… Ⅱ．①佳… Ⅲ．①饭店－建筑设计 Ⅳ．①TU247.4

中国版本图书馆CIP数据核字(2016)第021231号

创福美图 佳图文化 主编

中国林业出版社·建筑家居出版分社
责任编辑：李 顺 唐 杨
出版咨询：（010）83143569

出 版：中国林业出版社（100009 北京西城区德内大街刘海胡同7号）
网 站：http://lycb.forestry.gov.cn/
印 刷：利丰雅高印刷（深圳）有限公司
发 行：中国林业出版社
电 话：（010）83143500
版 次：2016年3月第1版
印 次：2016年3月第1次
开 本：889mm×1194mm 1/16
印 张：21
字 数：200千字
定 价：358.00元

Design Consultants
王启贤设计事务所

Starting in the early 80s, Mr. Clyde Wong worked for two internationally famous interior design firms, namely HBA (Hirsch Bedner & Associates) and DKA (Dale Keller Associates), and gained invaluable experience over the years by participating in projects, such as five-star hotels, clubhouses and villas. In 1993, he set up Wong's Interior and Associates Limited and subsequently WIA Design Consultants to undertake various kinds of projects covering interior designs and project management. In decades, WIA has accomplished numerous projects in Hong Kong as well as in the PRC; in addition, Mr. Wong was once appointed to enhance the new image design for the Bank of China (HK) Group and to act as the design Consultant to the Bank of Communications. WIA has received several awards including the Asia Pacific Interior Design Award, the Modern Decoration International Media Award and the Perspective Award. For the style of design, Mr. Wong incorporates different design concepts, e.g. simplicity, elegance, glamour, luxury, magnificence, avant-garde and classicism. His designs, especially those for villas, were very much approved by his clients and he was once given a title as 'the Magnificent Villa Designer'. In 2008, Clyde Wong Collection, the first book was published and distributed worldwide; and in 2016, Premier Quality Living is the second book featuring his works in the past few years.

由80年代开始，王启贤先生曾先后任职于两间世界著名设计顾问公司 - HBA (HIRSCH BEDNER & ASSOCIATES) 及 DKA (DALE KELLER & ASSOCIATES)，完成数项五星级酒店、会所及别墅设计、成绩斐然。1993年开始创立启贤建设有限公司和王启贤设计事务所，设计专业项目并提供设计工程管理服务，曾负责香港中银集团新印象设计及交通银行（香港总行）设计顾问，承接项目众多且广受好评，是香港资深的室内设计大家。经多年努力，获得包括亚太室内设计大奖、透视室内设计大奖在内的多个专业奖项。其设计风格华丽、典雅大方，深受业主喜爱，尤其擅长豪宅别墅设计和会所设计，有"豪宅王"的雅称。2008年出版第一本室内设计专集《居停集》，本书为2016年出版第二本专集《优享之旅·豪宅会所》。

香港理工大学设计硕士　　王启贤 Clyde Wong
M Des

CONTENTS

CLUBHOUSE & RESORT

006	Sky Wonderland Clubhouse in the Air		056	Enduring Fascination Luxurious Resort
024	A Spectrum in a Convivial Atmosphere Premier Clubhouse		082	European Luxury Corporate Clubhouse
044	Magnificent Performance Private Resort			

RESIDENTIAL

100	Sunlit Interior Villa of Simplicity		170	Nouveau Style Duplex 1
118	A Neoteric Space for the Post-80s Delicate Mock-up Room (50 sm)		180	A New / Classical Spirit Villa d' Art
126	Classy Inspiration Golf Chateau		192	Winning Combination Duplex 2
146	Contemporary Tempo Villa Moderne		200	Glamour and Luxury Villa in European Style
160	Modern Glamour Penthouse			

RESTAURANT & PUBLIC

210	Circular Tour Fusion Restaurant		236	Geometrical Extension Residential Buildings Public Area
220	Celestial Enchantment Exquisite Restaurant		244	Neo-Luxurious Expression Residential Lobby
226	Future Expectation Properties Sales Centre			

HOTEL

252	Curved Essence Chic Business Hotel		280	Modern Classic Business Hotel
260	Linear Perspective Hotel in Simple European Style			

COMERCIAL & LEISURE

296	Kaleidoscopic Silhouette Karaoke		314	Simplicity & Elegance Commercial Office
306	Abstract Expression Modern Office		322	Meditation Cottage Sanji Temple

SKY WONDERLAND

Clubhouse in the Air

The private club is situated at the top floor of a building. The snow white floating sculpture of the walls represents the clouds, and the streamline ceiling lights imitate the infinite orbit movements in the universe. Irregular-shaped lightings on the wall and the ceiling projected lightings resemble the stars twinkling in the far distance. The combined pattern of land and water current gives a bird-eye view of lands and oceans of the earth. The green growing plants on the walls indicates liveliness. All these elements are combined together to form a new and fresh space, a place full of life - Sky Wonderland.

Books are arranged in the bookcase on the snow white floating sculptured wall. It gives an impression of a library amid white clouds, forming a peaceful and silent reading environment. The library, the conference room and the refreshment area are well-linked in harmony with one another; and yet can be used independently. The streamline patterns of the ceiling and the floor, which are like road signs, direct people to different areas, e.g. the dining room and the suites.

The transparent leaf-patterned doors of the VIP room lead the guests into the room to explore another sky feature.

The VIP room is an extension of the starry space. White clouds and the star rays were painted on the wall and ceiling without obvious boundaries to create an endless horizon.

The outer space weaved with green leaves is part of the suite as they complement each other. Sleeping in the room is like taking a nap in the open sky. The green outdoor plants make the suite more lively and fresh. The main design concept was the change of space which would connect the indoor suite and outdoor roof area together in a hamonious manner.

The main theme of this private club house is "Sky Wonderland". How possible can it be, or can it be viewed from a different perspective?

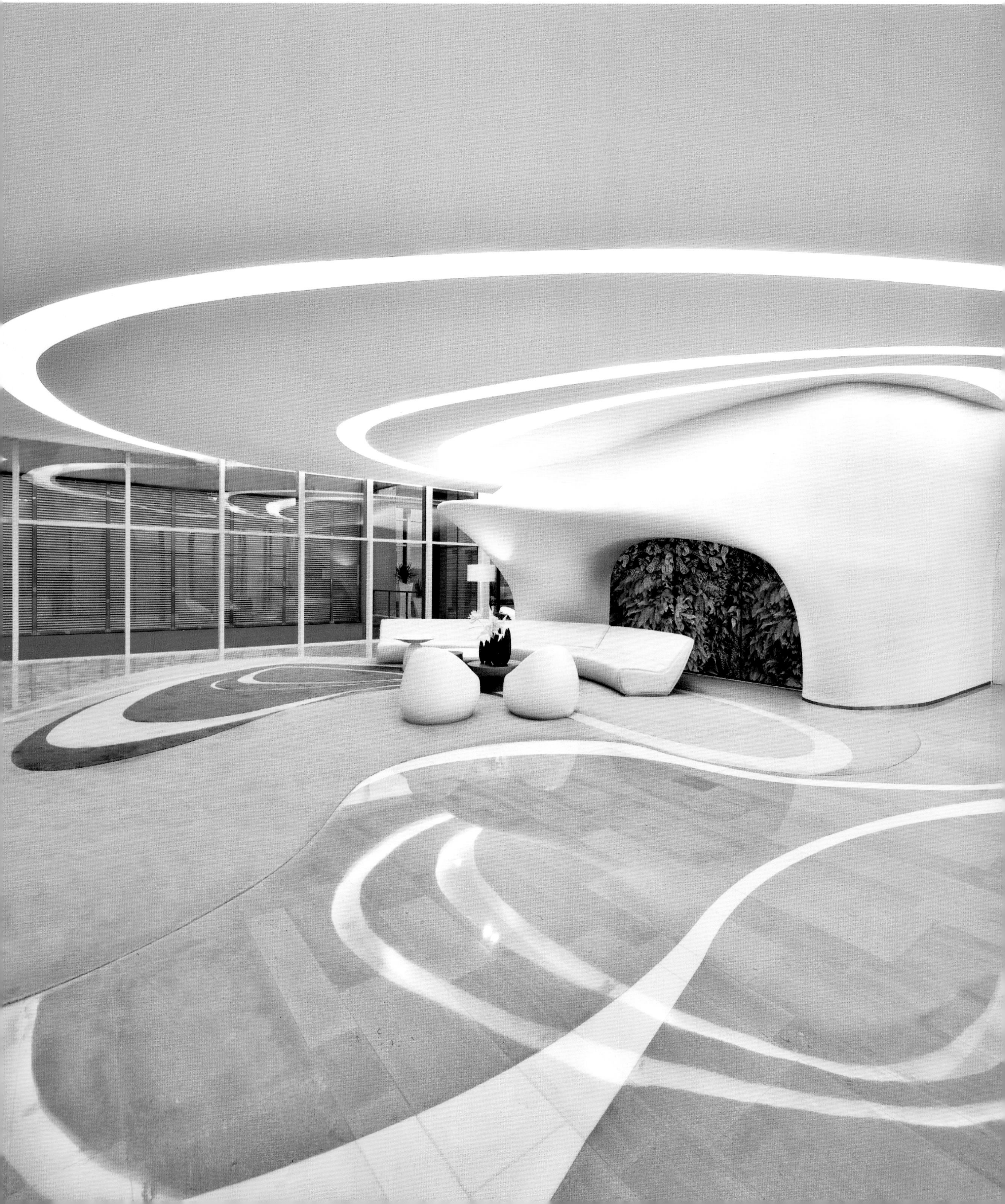

楼顶会所

天空之城

设计位于顶楼加建层的私人会所，雪白墙身浮雕代表白云；流线型灯槽天花表示天际星球自转轨迹；不规划大小形状的灯吼在墙身及天花投射光源，尤如星星般闪烁；泥土颜色与水流图案则是地球地面的展示；绿色植物墙身具有生命力的延续。将以上设计元素组合出一个新的空间，构成洁白而富生命力的天空之城。

将书籍摆设在雪白墙身浮雕中，像云层中的图书馆，带出安静的阅读环境，配合会议室及品茶区的联系，将三者溶为一体的活动空间。流线形天花及地面设计，把功能如水流带往不同的区域，一直延伸至用餐区及套房，如无形的指示路线图。

指向VIP房目标是树叶造型图案的双掩门，因通透光线的效果，引领客人推进房门，探索另一天空景界。

VIP房是星际的衍生延展，白色云彩与星星光线接连到天顶，间隔亦以放射的光线引伸至上空，没有明确的划分，打造无形界线的视野。绿叶编织的中空位与套房连成一体，睡在房中尤如在天空中无拘无束的自在，加上绿叶的户外植物，使空间更具生命力，套房的主线在于建筑空间的变化，将室外内的环境串连一起。

这私人会所的主题为天空之城，或者可用另一生活角度去审视它。

1. Entrance 入口
2. Leisure Area 休闲区
3. Reading Area 阅读区
4. Balcony 阳台
5. Conference Room 会议室
6. Teahouse 茶室
7. VIP Room 1 贵宾室 1
8. VIP Room 2 贵宾室 2
9. Kitchen 厨房
10. Garden 花园
11. Hallway 走廊
12. Suite 套房

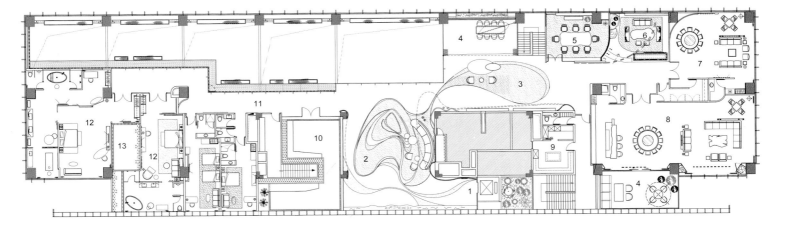

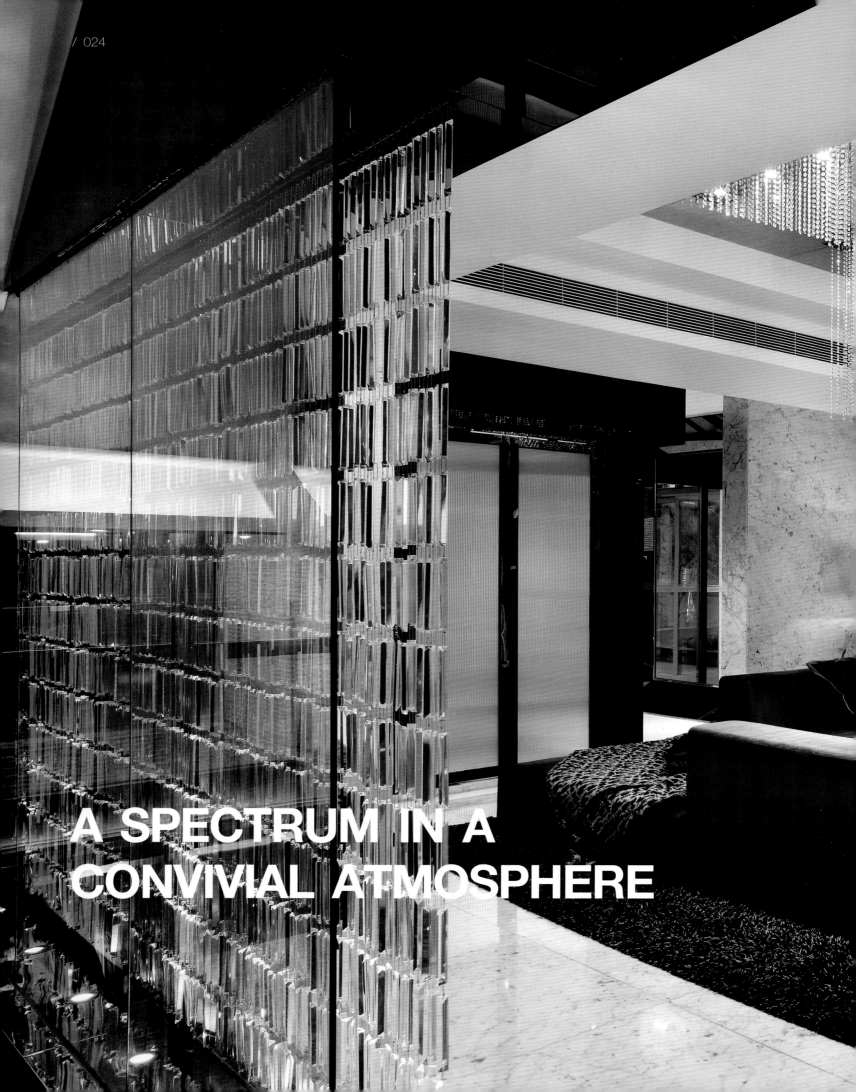

Premier Clubhouse

The villa is a 5-storey building of 800 square meters and was designed to become a fashionable holiday resort. A simple and yet elegant style has been adopted to construct a deluxe mansion, chic and not customary.

The main colour tone is grey as presented in the choice of marbles, and with silver paints, black glass, chandelier and other glistening decorative items make a balance between cold and warm colours. Further, fine lines with the geometric shaped concept in the ceiling present an extraordinary feature, e.g. the irregular-lined wine display units, stainless steel patterned panels and the large sofa set.

The upper level was designed as an ideal place for golfers to relax after the games. The ceiling is a sloping roof and three step layers built to form a large bordered frame for the magnificent scenery of the golf course and the lake. The large window lets in daylight enhancing the peacefulness with ease and comfort.

In the lower floors, the most important element is the different kinds of seating arrangement in the open area, e.g. sofa set, bar stool set, armchair set, platform bench set and others. Taking different seats, visitors may interact closely with friends in their own quiet and comfortable zone.

卓越会所

银线境峰

这所楼高五层的 800 平方米别墅成为客人闲时享受高品味度假会所，并以素净优雅的处理手法重新定义了 21 世纪时尚豪华的设计哲学。

本案例之设计以黑、白、银为基调，物料方面以白云石为主色调，与银漆面、黑玻璃、造型水晶吊灯及闪烁的装饰品互动地营造出高贵的质感。再加上柔和的黄色暖光，配以紫色为主的家具，成功地柔化了素冷的颜色基调。此外，适量的轻柔线条，配合天花的几何形状概念，成为了整个设计中具独特性的空间元素，就如在设计中出现的不规则线条酒架屏风、不锈钢图案屏风及大型梳化，其玩味令整栋会所增添了一份型格意象。

静态的顶层，善用斜屋顶的特点，做出层次踏级渗光天花造型，贯穿空间两端墙身，仿如平湖在画框内，恬静的气氛增添一份安逸，是打球者最佳松驰神经的落脚点。

动态的底层，最重要为座位布局，不同类型的雅座、梳化大组合、酒吧椅区域、4 人椅小组合、平台云石高低阶梯等，在开放形的空间设计，大家紧密函接而各有自己的角落。

/ 042

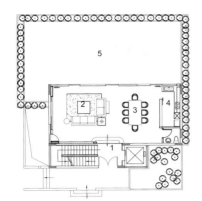

1F

1. Entrance 入口
2. Living Room 客厅
3. Dining Room 饭厅
4. Kitchen 厨房
5. Balcony 阳台

2F

1. Foyer 玄关
2. Suite 套房
3. Master Room 主人房
4. Study Room 书房
5. Balcony 阳台

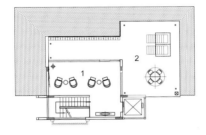

3F

1. Leisure Area 休闲间
2. Terrace 平台屋顶

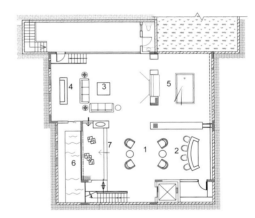

-2F

1. Leisure Area 休闲间
2. Teppanyaki 铁板烧
3. Living Room 客厅
4. Bar 酒吧
5. Snooker Room 桌球室
6. Shallow Pool 浅水池
7. Platform 高低地台

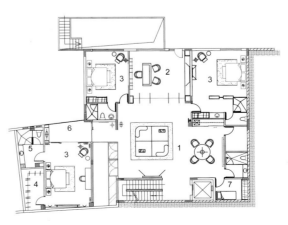

-1F

1. Leisure Area 休闲间
2. Study Room 书房
3. Suite 套房
4. Dressing Room 衣帽间
5. Bathroom 卫生间
6. Balcony 阳台
7. Servant Room 工人房

MAGNIFICENT PERFORMANCE

Private Resort

The designer adopted a mixture of noble, elegant, modern and extravagant style to create an enchanting and comfortable clubhouse.

Beyond the foyer, the ceiling lighting of glistening crystal acts together with the delicate European furniture for a simple and elegant ambience. The front of the staircase features an artwork screen matching with the diagonal patterned marble flooring and a large European classic painting.

The ceiling of the basements was left hollow to allow the display of the crystal chandeliers, which may enlarge the sense of height. The arc shaped panels and floor patterns form an extension, which lead the way to the small living room.

In the dinner room, the ceiling was constructed of horizontal false beams in stainless steel of champagne colour. The open kitchen with bar seats was built next to the dining room.

In the Karaoke room the glistening ceiling is the centre of attractions. The wall at the back of the sofa features specially decorated wallpaper and a black stainless steel shelving at the centre. The symmetrical chandeliers at the ceiling reflect its light on the coffee table creating a sense of interaction between the ceiling and the floor.

The overall layout of space, mixed with sculptures, paintings, designer furniture and lights, mimics the atmosphere of an art gallery, and it allows its resident to relax and enjoy the moment of tranquillity.

私人度假屋

瑰丽气派

设计以尊贵、典雅、现代、奢华的混和设计风格，将这幢楼高五层的别墅打造成一个让人陶醉的休闲会所。

从玄关踏进室内，即见天花上装置闪烁水晶的灯饰，配合精致的简欧家具衬托出典雅气韵。楼梯前的艺术玻璃画屏风设计，配合地台斜角的云石图案，再配合大型欧陆古式建筑挂画，每个细节都融合协调。

底层天花贯穿两层中空，有不规则排列的水晶工艺吊灯，高低造型扩阔高度感。弧形屏风与地台图案的延续，形态上引领客人步入休闲厅。饭厅天花使用排山型香槟金镜钢横梁，营造出一种无尽蔓延的感觉。旁边的开放式厨房，加上酒吧椅的摆设，带来更多互动功能。

卡啦OK包房天花闪烁雅致是华贵的聚焦点，沙发背后整面墙壁以特色墙纸及中间黑镜钢饰架互相衬托，丰富墙身美感。天花上设有对称精致的水晶灯，反射在茶几上，形成光影缤纷天地互动。

在空间整体布局上，配合雕塑、挂画、家具、灯饰等融为一体，艺术的气息遍布整栋别墅，让人放下疲倦，享受当下。

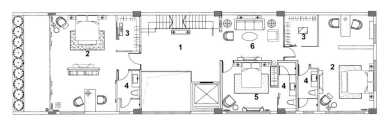

-1F

1. Hallway 走廊
2. Deluxe Suite 高级套房
3. Dressing Room 衣帽间
4. Bathroom 卫生间
5. Suite 套房
6. Leisure Area 休闲厅

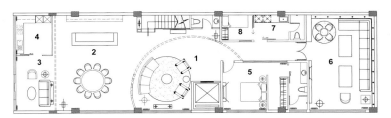

-2F

1. Living Room 客厅
2. Multi-Function Hall 多功能厅
3. Breakfast 早餐区
4. Kitchen 厨房
5. Suite 套房
6. KTV 卡啦 OK 房
7. Pantry 茶水间
8. Servant Room 工人房

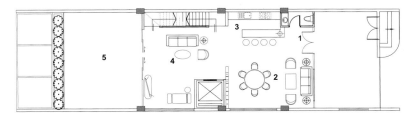

1F

1. Main Entrance 主入口
2. Dining Room 饭厅
3. Open Kitchen/Bar 开放式厨房 / 酒吧
4. Living Room 客厅
5. Balcony 阳台

2F

1. Leisure Room 休闲厅
2. Suite 套房
3. Dressing Room 衣帽间
4. Bathroom 卫生间
5. Balcony 阳台

3F

1. Study Room 阅读间
2. Storage Room 储藏室

Luxurious Resort

Covering an area of 1,700 square meters, this luxurious 5-storey resort villa consists of 14 guest rooms, and an assortment of function rooms, such as KTV room, multi-function room, VIP room, mahjong game room, and SPA rooms. The design followed the concept of European style.

In the lobby, the wall was covered with irregular plastic marble panels with recessed light troughs giving dim lighting effects. On the side, a wall was made of golden mirror with a large painting at the centre. The ceiling features a row of chandeliers.

In the multi-function room, a square-shape clear mirror was placed in the centre of the square-shape ceiling and was lined with hanging crystal lightings round the border. In the VIP room, the walls were covered with fabric panels, and the same design extended to that in the KTV room to establish a connection between the rooms. The ceiling in the KTV room features wooden carvings which consist of hidden LED lightings forming a continuous spectrum of colourful light waves.

In the basement, grey marble was used to cover the general flooring; whereas carpet was laid on the floor of the reading area to create a quiet environment. Books, reading materials and artistic decorations were placed on a row of full-height black lacquered shelving which also acts as a partition for privacy. The black lacquered shelving bears the similar design of the false beams in the ceiling forming cohesion.

The wall of the corridor leading to the suites was covered with classic bronze mirror to create a more spacious impression. In each suite, unique headboard was built to mark its distinctive features, which is unlike conventional hotels.

豪华渡假别墅

雅士迷醉 悠然大度

总面积达一千七百平方米，楼高五层內里有 14 间套房、KTV 房、多功能厅、VIP 包厢、麻雀房、SPA 房等等。以"欧式格调"为设计理念的豪华度假别墅。

走进会所的大堂中，背景墙以不规则云石晶片与暗灯槽错落编排，配合周边的金镜、巨型挂画，加上焦点性的天花吊灯，整体效果互相配合。VIP 包厢内，以大胆的红色软包墙身为主体。方形的天花，中央为方形清镜，以水晶围边吊灯，使天花层次分明。广阔开扬的多功能厅，适合举办大型发布会或红酒品尝宴会等等的多元化商业活动。VIP 房中，红色的布幕与 KTV 房的软包墙身形成一种连系，KTV 房天花上的木雕花，内藏 LED，七彩灯光的变化，带动 KTV 房的整体气氛。

在负一层，地台用灰色云石，而在阅读空间则铺放地毯，使阅读时更感到宁静，除了放置书籍及艺术摆设外，以通透的黑色高身层架为分隔屏障，增加客人阅读的私密空间。黑色层架与天花的假横梁形成一种连贯性的设计效果。前往套房通道的墙身，利用通透的古纹镜，减轻了走廊的压迫感。每间套房有不同的床头板特色设计，有别于一般千篇一律的酒店房间，更显出豪华精品度假别墅的特色。

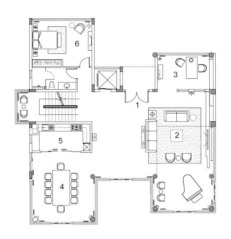

1F

1. Entrance 玄关
2. Living Room 客厅
3. Study Room 书房
4. Dining Room 饭厅
5. Kitchen 厨房
6. Suite 套房

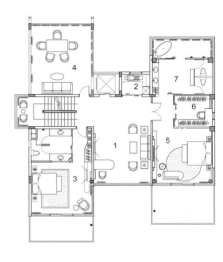

2F

1. Leisure Area 休闲厅
2. Pantry 茶水间
3. Suite 套房
4. Teahouse 茶室
5. Deluxe Suite 高级套房
6. Dressing Room 衣帽间
7. Bathroom 卫生间

HF

1. Leisure Area 休闲厅
2. Pantry 茶水间
3. Suite 套房

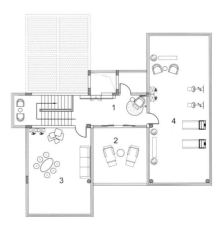

3F

1. Hallway 走廊
2. Balcony 阳台
3. Children Playhouse 小孩游戏室
4. GYM 健身房

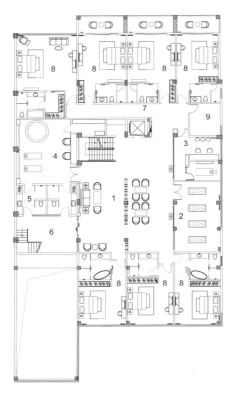

-1F

1. Reading Area 阅读间
2. Yoga Room 瑜伽室
3. Salon 美发间
4. Spa % Massage Room 水疗按摩间
5. Bathroom 卫生间
6. Balcony 阳台
7. Corridor 走廊
8. Suite 套房
9. Storage Room 储藏室

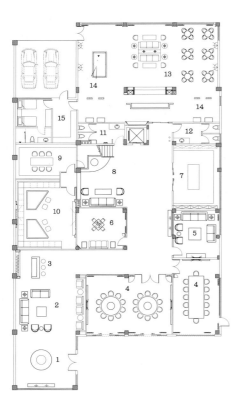

-2F

1. Entrance Lobby 入口大堂
2. Seating Area 座位
3. Bar 酒吧
4. Private Dining Room 私房菜包厢
5. Meeting Room 会客室
6. Mahjong Room 麻雀室
7. Kitchen 厨房
8. Hallway 走廊
9. Wine Cellar 红酒间
10. KTV 卡啦 OK 房
11. Male Toilet 男卫生间
12. Female Toilet 女卫生间
13. Multi-Function Room 多功能厅
14. Snooker & Darts Area 桌球 & 飞镖区
15. Staff Room 职员室

Corporate Clubhouse

Once in the 19th century in Europe popular neoclassical style Neo-Classicism. In the aesthetic characteristics of lies in its design style with European classical and modern design of the touch of the double effect of aesthetics, the perfect combination also allow people to enjoy the material civilization at the same time get spiritual comfort by. This concept in line with this case task. The goal of the design is to the approximately 1000 square meters, the building is 6 layer villa to become a group of senior club, and thereby allow the company's team after a long time fixes Bo after, enjoy a very superior class environment.

This case to have the European aristocracy and the same luxury taste and momentum, the emphasis is on the integrity, the founder of the pattern. From design ideas carried out in detail, inherited the neo classical former European palace relics of local gorgeous details, with beige, brown, golden warm tone, with black to sharpen the design effect, and added a certain proportion of white, with a proper warm and balance effect. Design in the process of discard too complex for the classical European style decoration simplified lines, with a simple European elegant feeling, place oneself among them, let company behind Kaiping, dense, enjoy the noble example of distinguished enjoyable holiday, reveals the unusual group of sakal, establish non general group status markers.

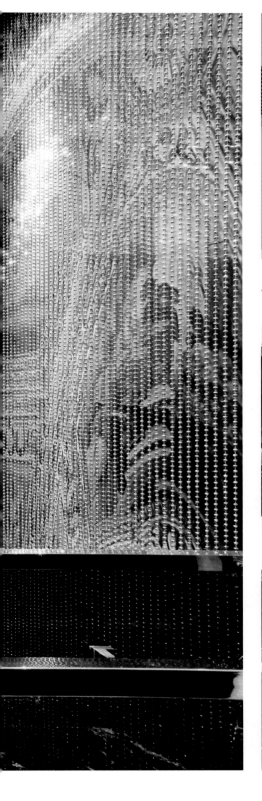

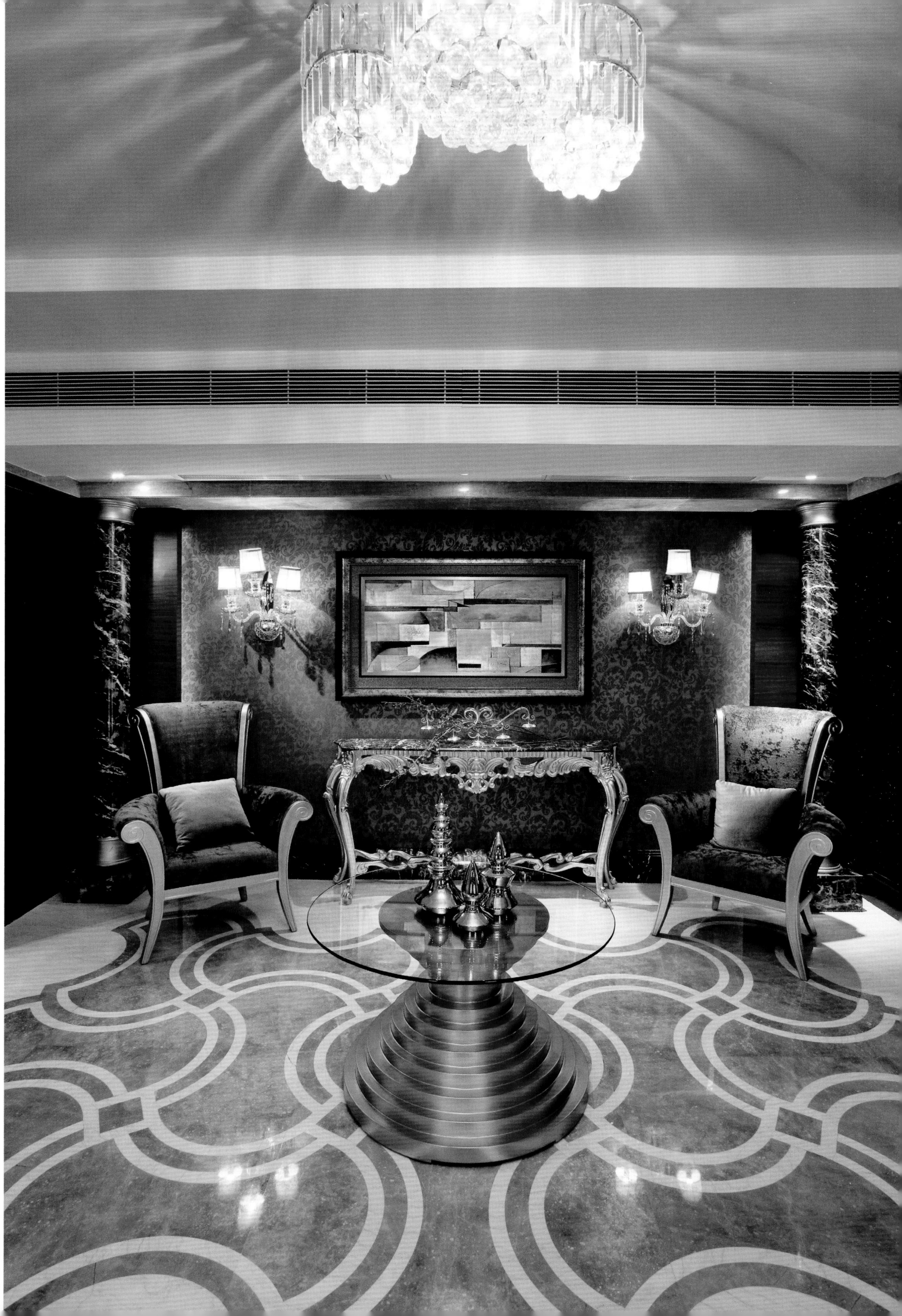

集团会所

欧丽豪庭

曾在 19 世纪于欧洲大行其道的新古典主义风格，在美学上的特质在于它的设计风格中具备欧陆古典与现代设计的触感产生的双重审美效果，这概念正切合本案例的任务。本设计的目标就是把这所约 1000 平方米，楼高 6 层的别墅成为一间集团公司的高级俱乐部，并借此让公司的团队们经过一番长时间拚博过后，尽情享受极级优越的环境。

本案例要有欧洲贵族同等的奢华品味与气势，从设计意念贯彻到细部，以米黄、啡、金的暖色为基调，配上黑色以锐化其设计效果，及添上一定比例的白色，带出适当的暖和及平衡效果。设计过程中弃掉过于复杂的古典欧式装饰，简化了线条，带出简欧典雅之感。置身其中，让公司上下抛开平日繁密的工作，享受高尚典范的尊贵写意假日，尽显集团超凡的卓见，确立非一般集团地位的标记。

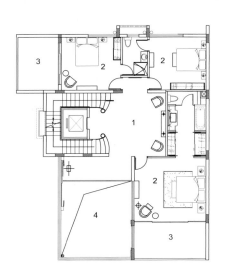

2F

1. Hallway 走廊
2. Suite 套房
3. Balcony 阳台
4. void 中空

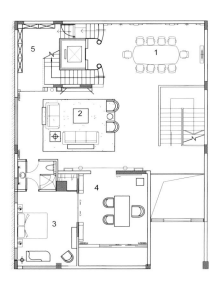

-1F

1. Dining Room 饭厅
2. Living Room 客厅
3. Suite 套房
4. Study Room 书房
5. Wine Cellar 红酒间

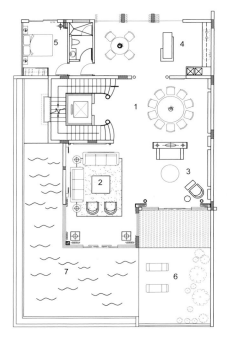

1F

1. Dining Area 用餐区
2. Living Room 客厅
3. Leisure Area 休闲区
4. Open Kitchen 开放式厨房
5. Suite 套房
6. Balcony 阳台
7. Swimming Pool 游泳池

-2F

1. Entrance 玄关
2. Theater 影音室
3. Servant Room 工人房
4. Garage 车库

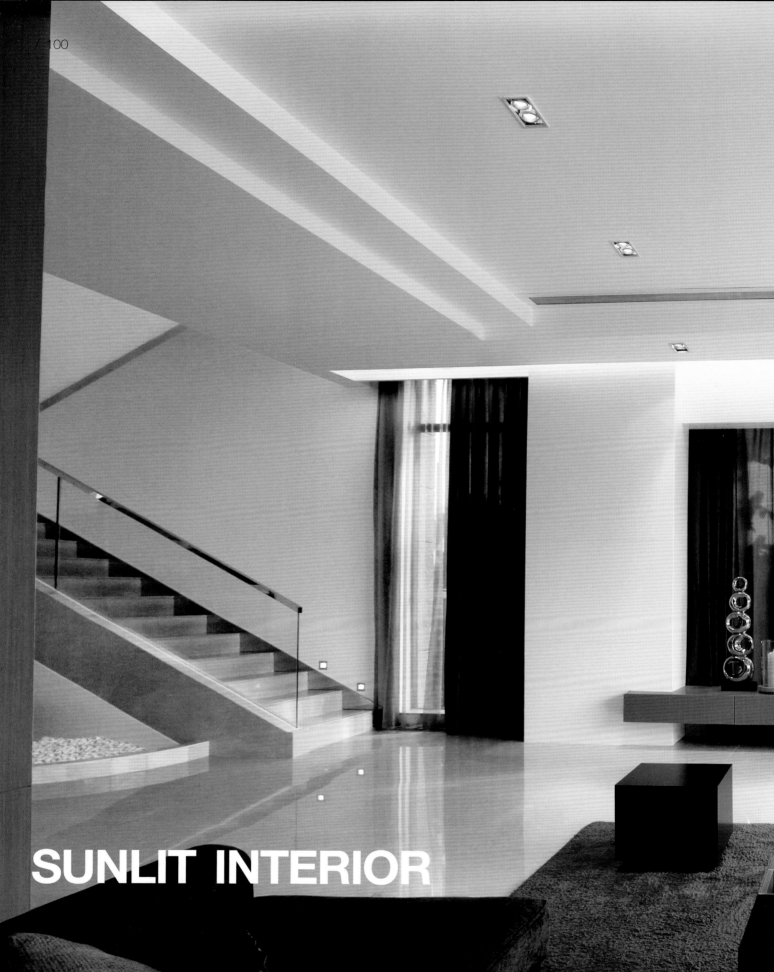
SUNLIT INTERIOR

Villa of Simplicity

The design aimed at creating a relaxed and tranquil place away from the hustle and bustle of city life. Closing the door to the garden will keep out the noise and pollution of the outside world. Around the centre of the villa, there is a 3-layer high ceiling window opening up to a idyllic view of natural and quiet scenery. The interior was made vivid with the all white and irregularly molded background.

The staircase was built with clear glass panels serving as a connection with the sky that may fill the interior space with natural daylight.

Downstairs, the karaoke room was designed in black mirror and oak wood. Adjacent to the karaoka room is the wine and chatting area. The two sections were partitioned off by a simply designed screen that made them appear to be identical in style.

The villa is situated in a row of town houses which poses a most serious problem, i.e. the lack of natural daylight. A large skylight was constructed to provide the interior with natural light; thus, helps reduce energy consumption.

The arrangement of each corner of each floor forms a continuation of clarity, and the overall shape displays elements of visual simplicity.

简约别墅
阳光府邸

当城市的财富创造者们将大部分的时间交给应酬饭局时,被压抑的神经渴望得到暂时的舒缓和宁静。这别墅就是挣脱城市紧张环境而设计。当我关上花园的大门,将城市的喧嚣和污染隔绝于外。环绕在别墅的中空,三层高度落地窗敞开一个自然幽静的空间,全白色的不规则凹凸背景造型令空间活泼生动起来。

沿着扶梯向上,穿过玻璃长廊,四周通透的玻璃设计,连通天与地,让自然光线贯穿整个室内空间,达至能源环保概念。

这时楼下传来爵士乐般松弛而专注的节拍,沿梯而下,在地库的卡啦OK房,以黑镜与原木的配搭,轻巧简单的线条将卡啦OK房与隔邻的红酒区清谈区分隔,两个开放式区域动线流畅,风格统一。

排屋的别墅建筑最令人烦恼是自然光线的缺乏,设计特意把屋顶天面打开,用透明玻璃覆盖,令自然光引入到全屋内,减小能源的消耗。

每层每角落的整体协调,形成一个干净俐落的连续,把简约时尚视觉的元素表露无遗。

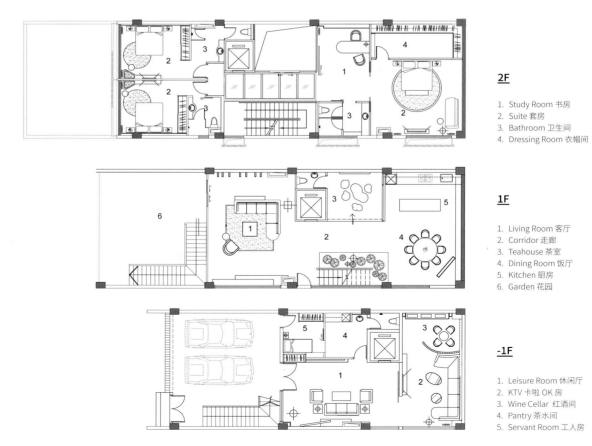

2F

1. Study Room 书房
2. Suite 套房
3. Bathroom 卫生间
4. Dressing Room 衣帽间

1F

1. Living Room 客厅
2. Corridor 走廊
3. Teahouse 茶室
4. Dining Room 饭厅
5. Kitchen 厨房
6. Garden 花园

-1F

1. Leisure Room 休闲厅
2. KTV 卡啦 OK 房
3. Wine Cellar 红酒间
4. Pantry 茶水间
5. Servant Room 工人房

3F

1. Skylight Space 天窗空间
2. Deluxe Suite 高级套房
3. Study Room 书房
4. Dressing Room 衣帽间
5. Storage Room 储藏室
6. Bathroom 卫生间
7. Balcony 阳台

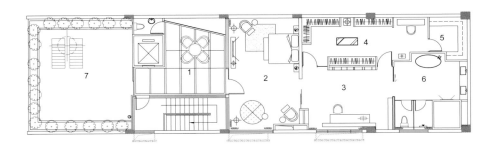

A NEOTERIC SPACE FOR THE POST-80S

Delicate Mock-up Room (50 sm)

A place of about 50 square meters but full of enchantment will let the Post-80s disregard the size, but to enjoy the interaction with each section that are cohesively linked together.

The Connection was carried out by the glass and white glossy paint, and together with the mirror they exert a dramatic visual impact - an illusion of wider space by transparency and reflections.

Plain white plays the main tune providing an interesting contrast with the partially black colored materials; and the upholstered dining chairs in red color add embellishment to the mono-colored environment.

The ceiling is a representation of the Milky Way with small LED lightings complemented by the mounted lights above the living and dining areas. The wall TV Screen and the irregular-shaped boxes form a dynamic object of contemporary art on the wall.

The design was intended to create a place of amazing comfort for the Post-80s, a boundless cyberspace within the boundary of the physical space.

精致样板房

时尚的 80's 后

打造 80 后年青一代的小居室，需体现一种空间魔术，使其富有迷幻感，趣味性和实用性，让来到这里的人们忘记它的大小，更多的沉浸在它的魅力和性格上，让空间自然的产生互动。

功能的分割以玻璃及白色光漆串联了每个区域，加大了视觉的快感。玻璃、镜面的常规运用制造出透明和反射，使空间更具有穿透力和迷幻性。

洁白的主色调与部分黑色质面形成冲突的对比。加上一点点红颜色餐椅的出现，使淡雅空间渗透出更有观赏性。

小型 LED 灯天花与餐厅吊灯结合，概念相等于宿影的星球及银河的外太空；挂墙电视与不规则的黑玻璃体结合，形成"流动的墙身艺术品"。约 50 平方米的居室，将现实的空间带出另一奇幻的舒适，满足八十后的简洁空间。

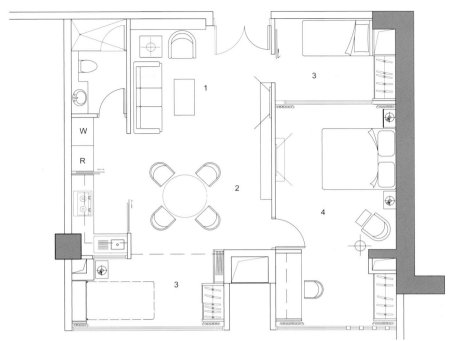

1. Living Room 客厅
2. Dining Room 饭厅
3. Suite 套房
4. Deluxe Suite 高级套房

CLASSY INSPIRATION

Golf Chateau

The 5-level villa consists of luxurious guest rooms, lobby, multi-function hall, bar, swimming pool, snooker room, karaoke room and banquet hall. The hotel apartments offer excellent facilities and are complemented by the lovely view over the lake.

The high ceiling in the lobby was decorated with crystal chandelier and sets of recessed square-shaped light trough. Together with the light-black coloured glass parapet in the mezzanine, they become a distinguish feature in the area. The design of the lobby of small hotel displays its magnitude in a delicate manner without excess and ostentation. This is precisely the style and rhythm that Yihe golf estate wishes to express.

Each floor holds 2 to 4 guest rooms, each with different decorative items. The floor was covered with natural textured wood. The delicately designed headboard acts like a piece of art on the wall. Laminated glass with patterns partition and the dark wooden cabinet complement each other to create a spacious ambiance.

The living and dinning areas were designed to provide a space without boundary. The kitchen was built next to the dinning area, and the 3 areas form together to fulfill logistic function. The full height shelving unit is covered with black glass and furnished with crystal decorative items inside. The window features roller blinds and marble frame. The 4 design selections bring out a sense of balance between different decorations.

新古典庄园
世纪光华

改造后共分五层,当中提供了豪华客房,大堂,多功能厅,酒吧,泳池,桌球室,卡拉OK及宴会室等。完善的别墅配套和室外绿柔柔景色,绝对是独一无二的珍贵宝玉。

大堂的高天花点缀了水晶球吊灯及方形的暗灯槽组合,再加上夹层走廊通道的浅黑色玻璃栏杆,两者上下连贯性的结合,成为该区的一大特色,设计彰显了中空优越的细致气势。

每楼层分别配置二至四个客房,每间房的设计确显不一样的摆设特色。富天然纹理的地板,特色床头板雕饰精致,像一件墙身艺术品的摆设,流水图案纹的夹心玻璃间隔与深色木板的衣柜,互相衬托,丰富了墙身质感。

客饭厅设计特征是通爽,室内外并无明显界限,空间序列流畅,饭厅和厨房之间划分清晰,三者共用同一宽敞的起居功能空间。黑玻璃高身层架配置水晶饰物,啡色布料,罗马帘与云石龙门窗框,四者刚柔并重,色彩分明,突显出丰富多样的物料材质,在细致的款式对比与不刻意的艺术饰物品之间带来和谐平衡。

即使是最朴实的人,也会不由自主地沉溺于美感的喜悦中。

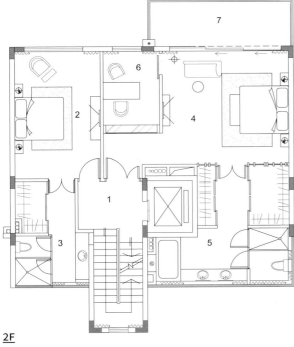

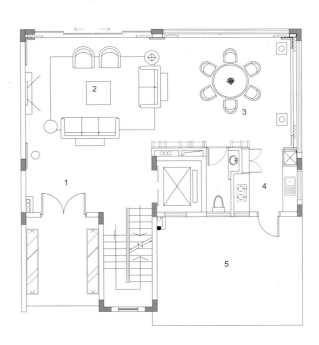

2F

1. Corridor 走廊
2. Suite 套房
3. Bathroom 卫生间
4. Master Room 主人房
5. Master Bathtoom 主卫生间
6. Study Room 书房
7. Balcony 阳台

1F

1. Main Entrance 主入口
2. Living Room 客厅
3. Dinging Room 饭厅
4. kitchen 厨房
5. Garage 车库
6. Toilet 卫生间

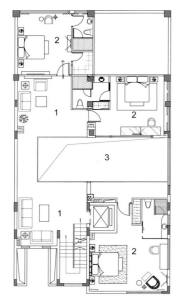

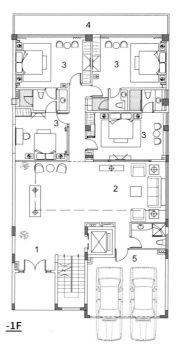

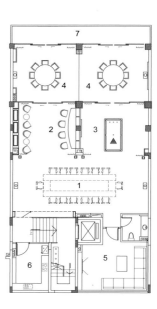

HF

1. Leisure Area 休闲区
2. Suite 套房
3. void 中空

-1F

1. Entrance 玄关
2. Living Room 客厅
3. Suite 套房
4. Balcony 阳台
5. Garage 车库

-2F

1. Multi-function Hall 多功能厅
2. Wine Bar 酒吧
3. Snooker Room 桌球室
4. Private Dining Room 私房菜包厢
5. Theater 影音室
6. Kitchen 厨房
7. Balcony 阳台

CONTEMPORARY TEMPO

Villa Moderne

The villa consists of 5 levels covering about 950 square meters. The design concept was to create a chic and relaxed ambience for the place; a design that may remain fashionable at all times.

Sahara beige yellow was adopted as the basis colour, accompanied by light colour wooden flooring and milky white materials as secondary colour throughout the villa. Other features, such as layered false ceiling, glassed flooring, walls of uneven surface, light well and raised floor, were used to give the impression of a contemporary lifestyle.

当代别墅

时尚旋律

当生活享受的定义早已被一般大众冠以一套像是金科玉律，且价值物质化的标准是法则时，精神上的深层品味往往会因此被忽略了。这所楼分五层，面积约950平方米的别墅正好造就了一次机会，于本案设计中演活了不造作中却展现出永恒不衰之魅力的品味生活。

这是一个简洁、明快的空间规划，时尚平和气息是这别墅的基本设计概念，更重要是注入一个"耐看"而不容易"过时"的作品。

建筑空间平面方案重新塑造后，功能上令人有耳目一新的感觉，造就了往后的光彩得以发挥。

材料以撒哈拉米黄作为色彩基调，配合浅色木面及奶白色材质渗透全幢别墅，跌级天花、玻璃地台、凹凸墙身、彩光天井及高低地台之造型令整体空间酝酿出极致的时尚生活品味。

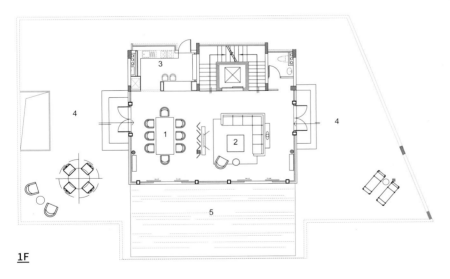

1F

1. Dining Room 饭厅
2. Living Room 客厅
3. Kitchen 厨房
4. Balcony 阳台
5. Swimming Pool 游泳池

2F

1. Corridor 走廊
2. Study Room 书房
3. Suite 套房
4. Dressing Room 衣帽间
5. Bathroom 卫生间

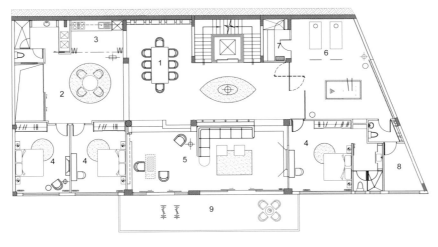

-1F

1. Dining Room 餐厅区
2. Breakfast 早餐区
3. Kitchen 厨房
4. Suite 套房
5. Reading Area 阅读间
6. Massage Room 按摩间
7. Sauna Room 桑拿室
8. Storage Room 储藏室
9. Balcony 阳台

-2F

1. Entrance 玄关
2. Leisure Area 休闲厅
3. Kitchen 厨房
4. Suite 套房
5. Void Garden 中空花园
6. KTV 卡拉 OK 房
7. Servant Room 工人房
8. Garage 车房

MODERN GLAMOUR

Penthouse

With a 4.5m high ceiling, the living room is spacious and bright facing spectacular scenery, the Swan Lake. Adjacent to the living room, the foyer is on a higher level and a balcony extension was built on its left with glass parapet. Standing there and staring into the living room, one may feel the enchanting views of the Swan Lake mingling with the interior, giving a sense of ease and comfort.

Modern glamour was the aim. Beige represents the basic color complemented by Rajah Teakwood and grey/black glasses. The structure depends mainly on lines and details. They interact with each other in this spacious realm, e.g. the square-form false ceiling, vertical recessed pattern walls, and the stainless steel inlaid marble floor.

The other four rooms bear the same tune; but with different details. High quality finishing expresses distinctive features in each room. Pure white marble, silver platinum panels, half-glossed wall paint and others coordinate well with each other. The Stylistic architectural fixtures complemented by the loose-piece display items exhibit a modern and glamorous residence.

顶楼公寓

现代呈现

四米半楼高客厅面对室外的人工天鹅湖，景致令人难以忘怀。半高的错层入口玄关左方，特意设计一个飘台，加上透明的玻璃栏杆，站立在此往外看，清雅的客厅好像与湖水内外交融，感受到额外舒适安逸，实在是非一般的高层复式单位。

现代魅力是设计意念的目标，米白色为本，深铁刀木及灰黑玻璃的配搭，结构着重线条及细节，譬如方框型天花架、直条凹凸灰玻璃理纹墙身、幼身的钢条线索地面图案，在这宽敞的空间内大家相互协调呼应。

上下共四个房间，运用相同的基本主调，每种面料都展示其独特质感。纯白云石、银铂漆板、灰黑色玛瑙石、半暗光漆黎明珠等等，在精简细节元素结合下，时尚格调的主体配合雅致的摆设，展现了现代优雅魅力的居停。

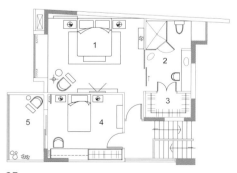

2F

1. Master Room 主人房
2. Bathroom 卫生间
3. Dressing Room 衣帽间
4. Bedroom 睡房
5. Balcony 阳台

1F

1. Entrance 玄关
2. Dining Room 饭厅
3. Theater 影音室
4. Kitchen 厨房
5. Living Room 客厅
6. Bedroom 睡房

NOUVEAU STYLE

Duplex 1

The distinctive feature of the project is the use of mirror-polished stainless steel which may display stronger reflection. White colour was adopted as the main theme, accompanied by warm coloured carpet and traditional European patterned furnishings.

The general design focused on details and its embellishment. White walls create a fresh and clear impression, and putting a large modern painting on the wall enriches the visual effect.

复式住宅 1
新派体现

本案带来了一种新的装饰理念。主要特点是采用光洁的反射度较强的镜钢材料。设计以白色为主调、其他只用作衬托。地毯用温馨的暖色调，欧式纹理混搭在家具和陈设中，这样的风格强调比例和色彩的和谐运用，线条简约流畅，尽显时尚现代气息。

设计注重细节的处理，客厅运用白色作为墙面的主饰面，试图打造出清新纯净的感觉，墙身搭配大型现代艺术挂画，有效丰富了空间的视觉效果。

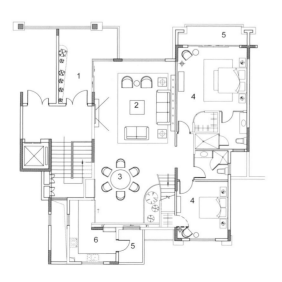

1F

1. Entrance 玄关
2. Living Room 客厅
3. Dining Room 饭厅
4. Suite 套房
5. Balcony 阳台
6. Kitchen 厨房

2F

1. Leisure Area 休闲厅
2. Study Room 书房
3. Suite 套房
4. Balcony 阳台
5. Deluxe Suite 高级套房
6. Bathroom 卫生间
7. Dressing Room 衣帽间

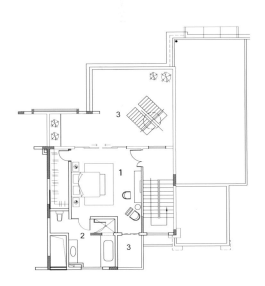

3F

1. Suite 套房
2. Bathroom 卫生间
3. Terrace 平台屋顶

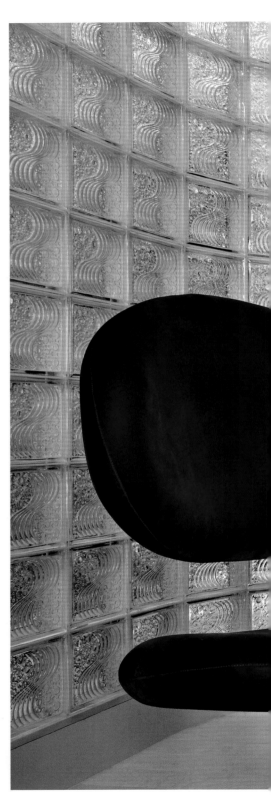

A NEW / CLASSICAL SPIRIT

Villa d' Art

The 5-storey villa incorporated a light European Classic design. Its facilities are customized to suit the needs of the residents who may enjoy every moment there.

The first 3 floors are equipped with general home facilities. To serve the needs of different routines, the master bedroom can be separated into 2 sections by a study room at the middle. If either spouse is working late, the partner can retire early in the separated room without disturbing each other. Basement 2 consists of multi-function rooms such as karaoke room, large dining hall and conference room. Basement 1 was constructed with 3 suites, living room, small pantry and servant quarter.

Fine furnishings are carefully selected, e.g. patterned carpet, light grey marble flooring, beige marble border and the crystal chandelier. Together, they reflect the style of the villa, i.e. nobility without ostentation.

雍雅别墅

古今风采

这座五层高的别墅融入了简欧的经典元素,高贵优雅,量身订造的个人化功能,缔造令人一再回味的奢华体验。

平面布局首三层是家庭配置。因照顾到不同客户的因素,设计上用创新的男女私人独立房间,中间共用书房,别具特式。负二层是多功能对外设施,有豪华卡啦OK室,大型用餐及会议厅。另负一层设三间套房、家庭厅、小茶水间及工人房。

瑰丽气派的用料搭配:图案地毯、浅灰云石地台、米色云石波打及上方水晶灯饰点缀的立体天花造型,充份展现尊荣感与内敛奢华的气派。

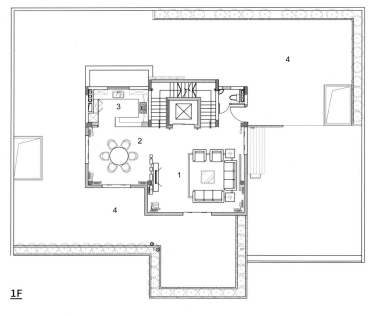

1F

1. Living Room 客厅
2. Dining Room 饭厅
3. Kitchen 厨房
4. Garden 花园

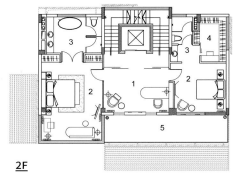

2F

1. Study Room 书房
2. Deluxe Suite 高级套房
3. Bathroom 卫生间
4. Dressing Room 衣帽间
5. Balcony 阳台

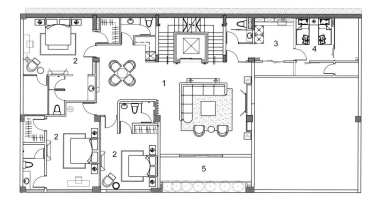

-1F

1. Living Room 客厅
2. Suite 套房
3. Pantry 茶水间
4. Servant Room 工人房
5. Balcony 阳台

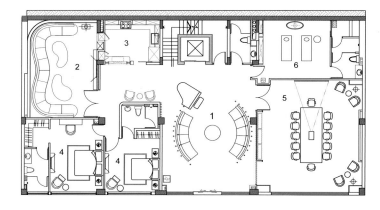

-2F

1. Piano Wine Bar 钢琴品酒吧
2. Theater 影音室
3. Kitchen 厨房
4. Suite 套房
5. Multi-Function Room 多功能厅
6. Massage Room 按摩室

Duplex 2

Artistic sculptures, abstract paintings, patterned carpets, veined marbles walls and flooring, distinctive lightings and spacious glass-partitioned bathroom are selected for this project. Together, they form a calm and relaxed ambience.

复式住宅 2

适度融合

艺术雕塑、形象挂画、图案地毯、云石纹理墙身与地台、独特的灯光及玻璃间隔的浴室,带给别墅一种沉静的韵味,在这个空间里完美融合。

装饰品凸显著居住者独特的审美情趣,个人生活感悟在细节中表现出来,而细节往往更显人们对美好家居生活的取向。巧妙的设计把居住者的生活品味、追求,融合在一起,不经意间的流露出来。

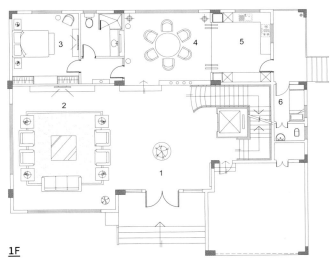

1F

1. Entrance 玄关
2. Living Room 客厅
3. Suite 套房
4. Dining Room 饭厅
5. Kitchen 厨房
6. Servant Room 工人房

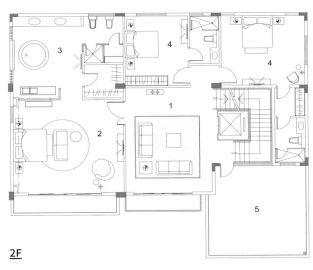

2F

1. Hallway 走廊
2. Deluxe Suite 高级套房
3. Bathroom 卫生间
4. Suite 套房
5. Balcony 阳台

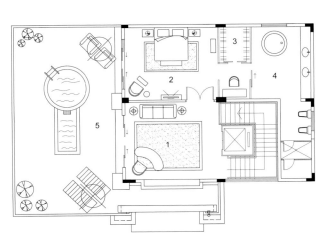

3F

1. Leisure Area 休闲厅
2. Deluxe Suite 高级套房
3. Dressing Room 衣帽间
4. Bathroom 卫生间
5. Balcony 阳台

GLAMOUR AND LUXURY

Villa in European style

Follow the lines of the design space layout, the taste of European style luxury style house.

First, open the door to see that light entrance on both sides of the seat, with a large sculpture placed in foil, create a majestic momentum.

Open living room in the broad space to form a visual cohesion, point out the rational layout of the building design. The main planning is change and division of architectural space, in order to extend the non barrier restaurant, the family hall, corridors and outdoor landscape.

Noble, elegant marble pillar distinguish, sitting room, restaurant and partial office space, smallpox by mirror design, European style furniture, carpet, red curtains, European clues decoration and art furnishings, and so on.

For some of the European style design of entrepreneurs, the ideal house is such a big. The spatial pattern of here stand stable, as in the European taste.

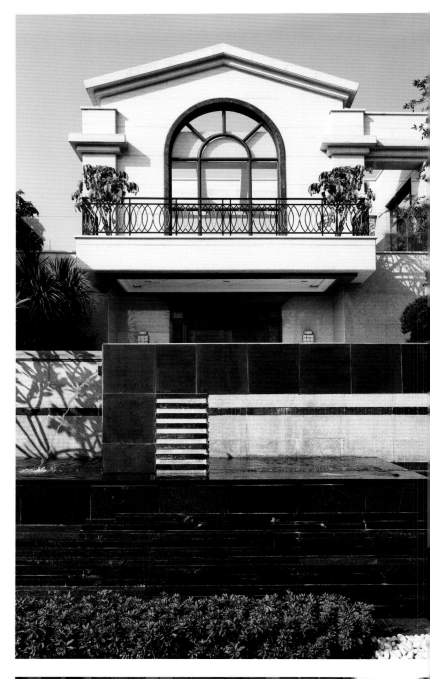

欧式别墅

艳丽奢华

设计依循着空间框架的布局，品味欧式大宅的奢华风格。

首先，开门即见玄关两侧的透光座，用大型雕塑放置衬托在上，营造磅礴气势。

开放式的客厅在空间中形成视觉中心，点出建筑设计上的合理布局。主规划在于建筑空间的改动及划分，以无阻隔的餐厅、家庭厅、回廊及户外景观的延伸。

高贵、优雅的大理石柱区分客厅、餐厅与偏厅的空间，天花采用镜面设计，配搭欧式家具，波斯地毯，鲜红窗帘布，欧式线索装饰及艺术摆设等等。

对于一些喜欢欧式设计的企业家而言，理想的住宅在空间格局上必需四平八稳，亦在这里体会了欧式的豪华品味。

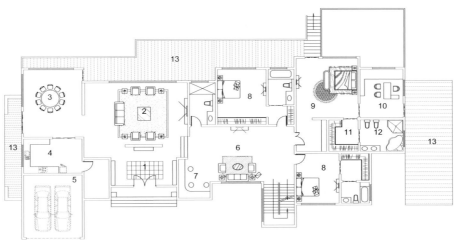

1F

1. Entrance 玄关
2. Living Room 客厅
3. Dining Room 饭厅
4. Kitchen 厨房
5. Garage 车库
6. Leisure Area 休闲厅
7. Bar 酒吧
8. Suite 套房
9. Deluxe Suite 高级套房
10. Study Room 书房
11. Dressing Room 衣帽间
12. Bathroom 卫生间
13. Balcony 阳台

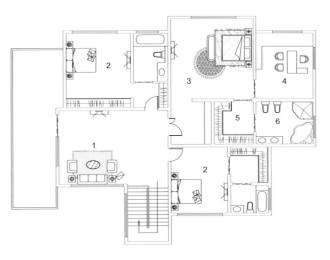

2F

1. Leisure Area 休闲厅
2. Suite 套房
3. Deluxe Suite 高级套房
4. Study Room 书房
5. Dressing Room 衣帽间
6. Bathroom 卫生间

Fusion Restaurant

The design theme was aimed at presenting the nature in figurative terms. The presentation begins with a circular feature in the centre as the focal point surrounded by a number of VIP rooms; then, the swimming pool of the clubhouse in the exterior. It is intended to create a calm and relaxed atmosphere for the guests to enjoy their time on an imaginary island.

The design concept was derived from Contemporary Design with the choice of colours in beige and light brown enhanced by black. The ceiling was constructed with white aluminium boards scattered with irregular leafed-shaped openings, projecting a leafy shade by small spotlights and strips of LED above. The lighting system appears as if some small stars and meteors in the night sky.

In the dinning hall, geometrical or clean-lines design was adopted: square-shaped dining tables with round base were placed according to the circular features around the hall, the geometrical-shaped columns stand in the centre with parapet of glass in between, V-shaped pattern of wooden floor, black beetle-shaped wall lamps on columns, and bright round bands of the lightings in the ceiling resemble the galaxy. They all help to enclose, to separate and to interact with each other in every corner of the hall.

In addition to the main dining area, the restaurant contains of two other sections: VIP rooms of western cuisine and VIP rooms of Chinese cuisine, which are drawn separately by the light brown coloured corridor. Large VIP rooms are positioned next to the swimming pool with a superb panorama of the view. Guests may arrange having a private party in the room using the pool and at the same time enjoying the excellent cuisine.

Through the concept of admiring the nature, the designer used different materials to translate the idea into reality. In a cold architectural environment, he adopted a new design concept to help express the feeling of the people towards the Mother Nature.

新派餐厅

天际回廊

餐厅设计以抽象的大自然景物为主题演绎，整个餐厅营造了圆形的地势，中间为核心点，外围以多间 VIP 餐房包围，外边为会馆泳池，像于圆形岛屿中用餐。

餐厅以现代时尚风格，室内整体以米黄，浅啡等作主调色，配以黑色点缀。天花采用抽象树叶形梳孔白色铝板，一些加置的小射灯及长弧形的 LED 灯条，两者看来像天上的流星及小星球，显出整个空间有如致身晚上的星空中。

大堂用餐区的餐桌跟随圆形地势排列得疏落有序，餐厅中心用白色几何形支柱，对空间进行围合分隔，有如半私人区域，配以半身玻璃围绕，使空间相互渗透。细节上采用 V 形排列的天然木地板，像甲虫外形的黑砂钢柱身壁灯，环绕地球的月亮圆形灯带，把一切融会一起，写入每个角落。

本餐厅共分两环，除了大堂主要用餐区，更有西餐及中餐包房。通过中性的米啡色走廊，把中西包房相互划分。大型房间更是设计在约 180 度环绕室外泳池的观景位置，客人品尝美食同时，可享用泳池与私人派对的互动功能。

透过大自然的实际概念，再转化用不同的物料元素，在混凝土冷冰冰的建筑框架内，用新的设计思维，表达了现代人希望拥有对大自然的感情。

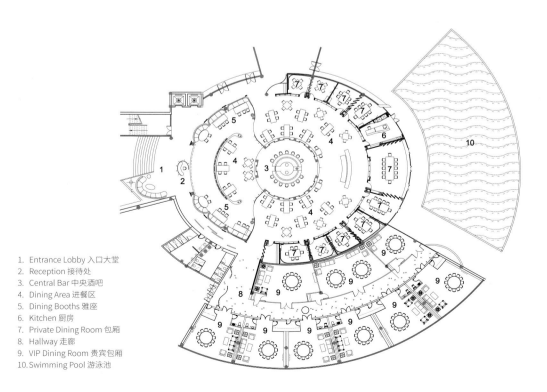

1. Entrance Lobby 入口大堂
2. Reception 接待处
3. Central Bar 中央酒吧
4. Dining Area 进餐区
5. Dining Booths 雅座
6. Kitchen 厨房
7. Private Dining Room 包厢
8. Hallway 走廊
9. VIP Dining Room 贵宾包厢
10. Swimming Pool 游泳池

/ 220

CELESTIAL ENCHANTMENT

Exquisite Restaurant

Abstract design was the main concept of the restaurant and the sky with stars and clouds were the main theme.

The entire restaurant was mainly built with three types of simple materials, i.e. wood, white beige paint and stone floorings. The walls were decorated with curved lines and recessed lightings giving an impression of fluffy moving clouds. The ceiling was constructed in winding layers connected with LED lightings. Together with the tiny spotlights, they display a starry, starry night.

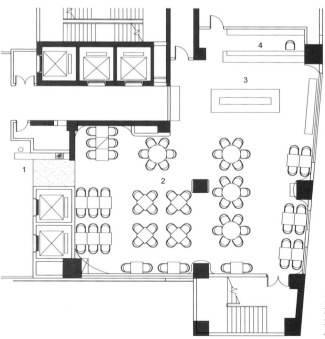

西餐厅

星空魅力

抽象设计为餐厅的主要概念,纯粹用设计手法打造星际、云层的主题。整个餐厅以三种简单材料为主,有木面、奶白油漆和石材地面,墙身的弧线加上灯光,好像起落飘浮的云彩,天花上层层叠起,弯弯曲曲,配合光带,加上星星设计的灯光,媲美夜空的天际。

1. Reception 接待处
2. Dining Area 用餐处
3. Buffet Bar 自助取餐吧
4. Soft Drink Bar 饮料吧

FUTURE EXPECTATION

Properties Sales Centre

The project is situated next to a wetland park. Red colour was chosen for the exterior to create a warm feel; while white colour was chosen as the main tone for the interior to express purity.

The interior decoration was planned to provide clear guidance to customers when entering the centre. To fulfill the function, the designer adopted lighting arrangement to separate different sections in a natural way.

The ceiling arrangement features curved lined design giving the effect of a light, chic and natural scene. Planters with bamboos are placed in some areas to imply the importance of a green environment. In Chinese culture, bamboos are generally regarded as having a pure and elegant character, and their presence may give the interior an air of elegance.

The round bar table was linked with a long flow of plastic cloth which looks like a ribbon flying towards the ceiling; thus creating a stronger impact.

The place exhibiting the housing models is the key factor of the project. Here, it was designed to facilitate customers to walk around and to obtain the required information easily and rapidly.

As a property sales center, it should provide a comfortable environment, in which the customers may look forward to their ideal lifestyle.

售楼中心
生活的憧憬

位于西安东北部浐灞生态核心区，本案作为销售中心项目，在整个格调上均与毗邻的湿地公园融合相连。室外红外墙的暖和感与室内白色主轴的纯净，是颜色运用的明确分工。

整个室内空间布局规划是给客户进入大堂后的明确指引；设计为了强调这种引导功能，在灯光的使用上，更加注重柔和的光源配置，让不同区域显得更加独立、自然。
吊顶排列的曲弧线造型设计，呈现出了一种柔和时尚效果，同时也更为轻盈自然，体现了天花美感。
部份区域打造室内花圃，一些种植竹子，在视觉上营造出植物在室内空间环境中的重要性，竹子所代表着清新脱俗的品格，也给空间带来了几分淡淡的清幽之气。设计对空间的流畅感把握要恰到好处，像圆形吧台与一个流线型的造型相连接，像丝带般在天空中飘荡，造成强烈的视觉冲击。

沙盘展示区是整个售楼中心关键所在，让客户在模型空间中可以最为直接地关注自己所需要的资讯。
作为售楼中心，必让客户得到了一份对未来美好生活的憧憬。

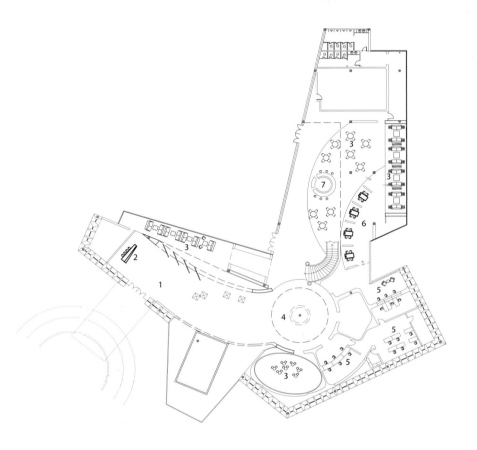

1. Entrance Lobby 入口大堂
2. Reception 接待处
3. Seating Area 座位
4. Huge Statue 巨形雕像
5. Office 办公室
6. Meeting Area 会客区
7. Snack Bar 小食吧

GEOMETRICAL EXTENSION

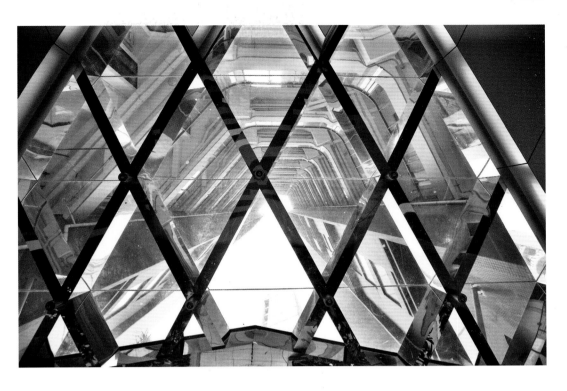

Residential Buildings
Public Area

The two residential buildings were designed to display modern architecture. The exterior walls of the building were embedded with random geometric patterns. They are hidden lines of vivid colours with different effects from different angles.

At the entrance, the ceiling is plainly the centre of attraction. It is a canopy built in glass and stainless steel as frame. It appears in a traditional diamond pattern with its base designed in an oval shape like a ring holding a diamond. For the walls and door, stainless steel strips were used with silver Blackstone to express a sense of a beautiful starry night. This mood is further enhanced by the floor which was constructed with different shades of grey stones.

The walls in the lift lobby were covered with Blackstone reflecting the dramatic landscape of the nature. Together with the soft lighting arrangement, it creates a formal and yet interesting atmosphere. At the entrance to the lift lobby, stainless steel strips were also used to provide a continuation of the main lobby. The recessed lighting effects create an interaction between the ceiling and the floor patterns, thus making the patterns more dimensional.

1. Ground Floor Ceiling 首层天花
2. Corridor 走廊
3. Lift Lobby 电梯等候区

一栋标准层电梯大堂

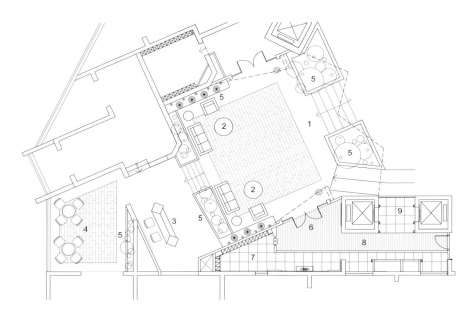

一栋入户大堂

1. Main Entrance 入口
2. Seating Area 等候区
3. Reception 接待处
4. Balcony 阳台
5. Plant 植物
6. Lobby Entrance 大堂入口
7. Mailbox 信箱
8. Corridor 走廊
9. Lift waiting Area 电梯等候处

住宅大厦公共区域

几何延展

两栋现代感十足的高级住宅大厦。外建筑以不规则编排的几何图案为设计主轴，用线条勾勒出层次，从不同的角度观察，会发现有鲜明的颜色隐藏其中，低调的展现不平凡。

两栋楼宇的入户大堂，最引人注目是天花的菱形玻璃棚，外形酷似一颗切割的钻石图案，就像承托钻石的指环。墙身与大门的镜钢条配搭黑色背景，尤如流星群滑过夜空，加上三种不同灰度的石材拼砌而成的地台，为住客提供了一种几何的视觉效果。

电梯大堂的墙身运用稳重的黑色石材，排山造型的天与地互相衬托。柔和的灯光为空间营造出庄重气氛。大门入口处同样采用镜钢条不规则排列装饰，让人意识到设计空间方向的延续。天花的渗光效果与地面的图案互相呼应，加强了立体感。

入户大堂的景观，道尽了设计对于几何图案及线条的玩味性，两者混合设计营造出新颖的感官体验。

/ 243

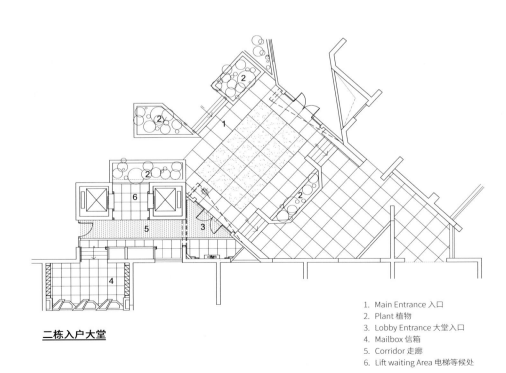

二栋入户大堂

1. Main Entrance 入口
2. Plant 植物
3. Lobby Entrance 大堂入口
4. Mailbox 信箱
5. Corridor 走廊
6. Lift waiting Area 电梯等候处

NEO-LUXURIOUS EXPRESSION

Residential Lobby

For the lobby of a prestigious building, the designer based on using geometric patterns with vertical crystal lightings to establish a majestic but not too solemn atmosphere.

The main focus of the design was the ceiling, where the vertical crystal lighting was hung from the center. At the lift lobby, the ceiling was covered with mirror-polished stainless steel, by which the walls appear to be higher than actual size.

The overall design depicts the use of the colour contrast of materials, such as light beige coloured flooring, and dark brown teakwood with black marbled walls. Together with other furnishings, such as geometric paintings, and patterned walls, they help strengthen the visual impression of the space.

公寓大堂

新派华丽演绎

作为高贵大厦的住宅接待大堂，室内设计以几何图形配合仿直水晶灯珠，提供给住客及来宾一份高贵而不过份隆重的舒适体验。

天花是重要的环节，将垂直水晶灯组合聚焦在主要中心位置，电堂大堂采用镜钢加强墙身的高度感。

色彩的演释更凸显出深浅的对比，浅米白色的地台，搭配深啡色的木面及稳重的黑麻石墙身，加上几何图案的艺术画作及图案墙身的布置，丰富了空间的视觉印象。

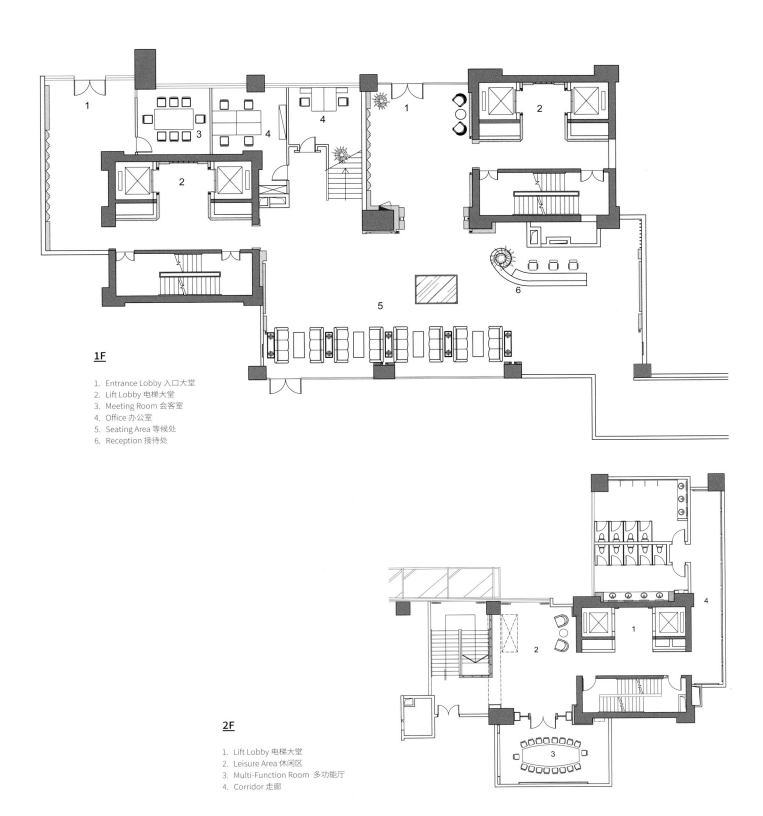

1F

1. Entrance Lobby 入口大堂
2. Lift Lobby 电梯大堂
3. Meeting Room 会客室
4. Office 办公室
5. Seating Area 等候处
6. Reception 接待处

2F

1. Lift Lobby 电梯大堂
2. Leisure Area 休闲区
3. Multi-Function Room 多功能厅
4. Corridor 走廊

/ 252

CURVED ESSENCE

Chic Business Hotel

Received a bump in the art of wood surface relief as the background. The reception counter arc design, platform three smallpox and irregular with each other, producing at different visual effects. Platform with different colors of marble pattern, smallpox parabolic lamp troughs and pits black line, changes in a different sense of fashion. The lobby of the surrounding with eight multi function hall, with the function of Traders Hotel.

Abstract design for the main concept of the restaurant, and wall body of the arc light, soft like clouds floating in the ups and downs, smallpox superstratum cascade, with a winding ribbon, and star design lights, the night sky is comparable to that of the sky. The whole restaurant with three simple materials, with wood, milk white paint and stone floor, pure design techniques to create a star, cloud theme.

时尚商务酒店

旋律动向伸展

接待处以凹凸的艺术木面浮雕为背景，地台用不同颜色的云石组成图案，天花有抛物线的灯槽及凹坑黑线，三者不规律的弧线形设计互相结合，变化出不同层次的视觉效果。大堂的周边设有八间多功能厅，配合商务酒店的功能。

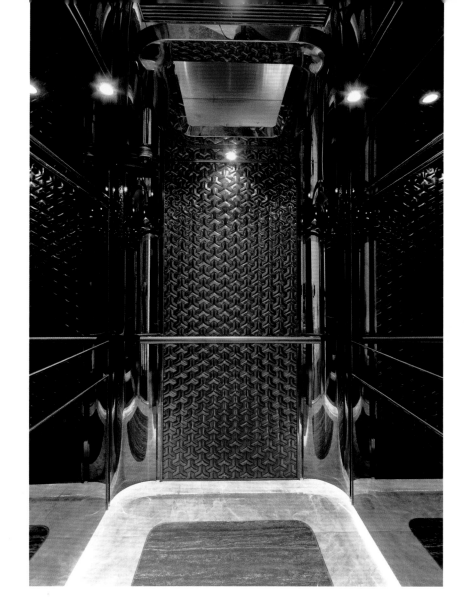
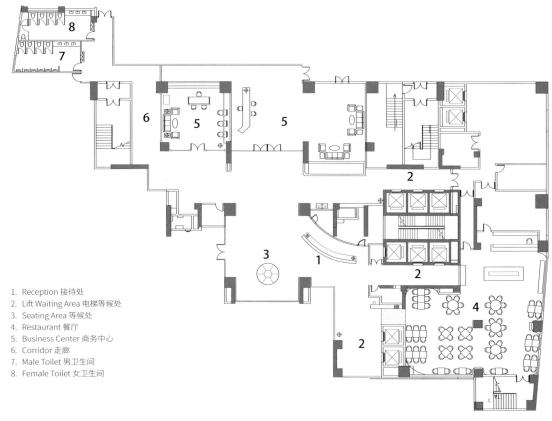

1. Reception 接待处
2. Lift Waiting Area 电梯等候处
3. Seating Area 等候处
4. Restaurant 餐厅
5. Business Center 商务中心
6. Corridor 走廊
7. Male Toilet 男卫生间
8. Female Toilet 女卫生间

Hotel in Simple European Style

At the entrance of the lobby, the tall bronze steel artwork immediately catches attention. It displays the artistic formation of lines and also acts as a partition for the bar and lobby area. In the reception area, two large lamps were placed on a base at each side. In the lobby, transparent glass was used as the flooring to visually open up the swimming pool below. This design renders the area of a sense of being afloat. Simple large cushions were selected to replace traditional seats.

In the simple European themed premium rooms, the main focus is the large partition frame separating the dining and living rooms. Living room features hard patterned marbles, whereas the bedroom soft carpets.

Located at the top is the presidential suite. The main focus was to leave open the flooring in the centre between the ground and upper floors to create a more spacious feel. The main floor consists of rooms made for gatherings, such as living room, dining room, meeting room, and kitchen and guest rooms; whereas the upper floor is mainly for bedrooms. In the suite, the bathroom is more spacious, in which a massaging hot tub was installed at the centre with a chandelier above it. On the two sides, basin counters for men and women are placed separately. In the rear, frosted glass partition with stainless steel framed pattern was set up as a room divider for shower and toilet facilities.

简欧酒店

绽放轮廓

进入大堂,焦点就在中央的高身通透古铜钢屏风,展现丰富线条的轮廓,亦将大堂与酒吧区作分隔用途,使两者间有隐约互动。接待柜台两旁的灯座,给人一种庄严款待的感觉。大堂的地台设计特色是采用了清玻璃板,打通了楼层,清透的玻璃可看到下层的泳池,悬空的感觉增加趣味性。大堂选用简洁的大圆形咕臣座家具,简化了传统的大堂座椅形式。

简欧设计风格的高级套房,焦点在于大型龙门框,分隔客厅与饭厅,客厅地台以硬身云石拼花为重心,睡房以软身地毯铺设,质感轻柔。

位于顶层的总统套房,重心在于打通上层的地台,营造一个中空位置,令空间更开扬广阔。而且使两层可互相呼应产生关连。首层以公用空间为主,有客厅、饭厅、会议室、厨房及随从室,而楼上以客房为主。总统套房的卫生间比例较宽敞,中央摆放一个焦点性的按摩浴缸及水晶吊灯在天花,左右两旁分别设置男女洗手柜台,后方有磨砂玻璃与钢框图案企浴及座厕间格,打造一个总统套房的豪华浴室。

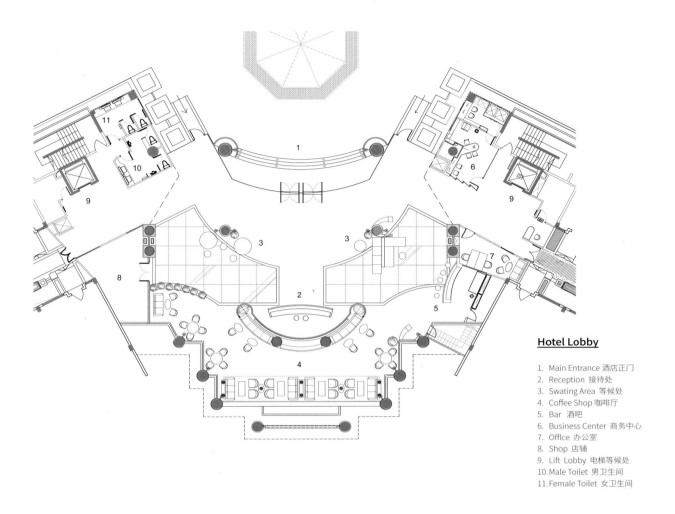

Hotel Lobby

1. Main Entrance 酒店正门
2. Reception 接待处
3. Swating Area 等候处
4. Coffee Shop 咖啡厅
5. Bar 酒吧
6. Business Center 商务中心
7. Offlce 办公室
8. Shop 店铺
9. Lift Lobby 电梯等候处
10. Male Toilet 男卫生间
11. Female Toilet 女卫生间

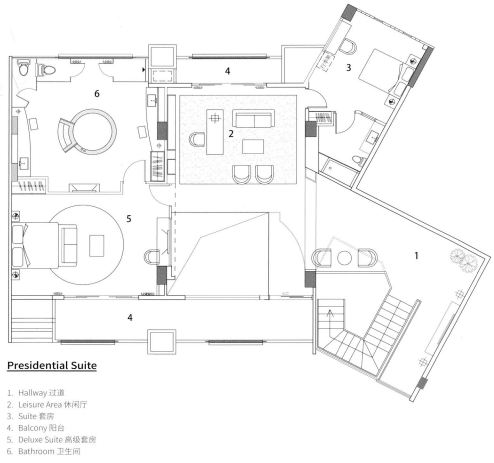

Presidential Suite

1. Hallway 过道
2. Leisure Area 休闲厅
3. Suite 套房
4. Balcony 阳台
5. Deluxe Suite 高级套房
6. Bathroom 卫生间

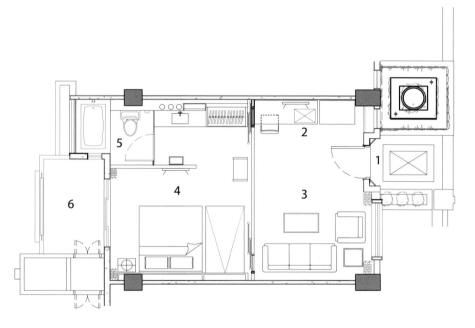

suite 1

1. Entrance 玄关
2. Desk & Television Set 电视组合柜台
3. Living Room 客厅
4. Bedroom 睡房
5. Bathroom 卫生间
6. Balcony 阳台

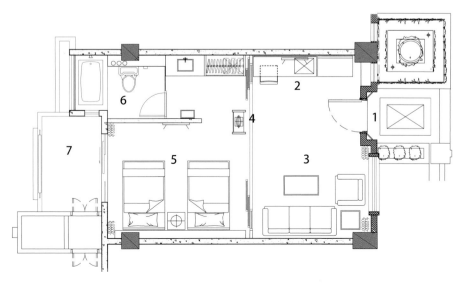

suite 2

1. Entrance 玄关
2. Desk & Television Set 电视组合柜台
3. Living Room 客厅
4. Bar 酒吧
5. Bedroom 睡房
6. Bathroom 卫生间
7. Balcony 阳台

MODERN CLASSIC

Business Hotel

The hotel entrance lobby is the focus of the ceiling star shaped cone pendant lamp, focus to the lobby of the central. In the marble column in exquisite sculpture with a white frame, on the combination of two layers, like a large wall reliefs. The pattern of wave shaped lead guests come to the reception counter. The reception counter by printing aluminum fabric and irregular dark light plain wall to slash, full of artistic temperament. The central hall placed a noble round seat, with the formation of a floral art decoration. A form of smallpox resembles a blossoming roses bloom. The bar area of the design and the reception counter wall echo each other, half of the screen bar area and lobby separated, with privacy and will not feel cramped. West dining room at the entrance of the gray glass mirror steel with irregular curved screen, the use of the material of the reflection characteristics, add perspective. Unsightly building scape, wrapped in aluminum bronze into oval shaped, the Fust naturally into the space, is also indicating a main entrance. Restaurant milk white ceilings, design to the drops of water ripples patterns as the center, is arranged below a buffet, with heaven and earth, making it one of the restaurant's activity center. Came to the meeting room, just like walking into a box and smallpox central mirror steel with on both sides of the decorative plate with pattern, like Tianwaiyoutian space scenes, and platform of arc pattern and echo each other, make visual effect more ultra three-dimensional sense of space.

The two layer is the open leisure center, reception rooms also contain. Reception at the concept in an arc mainly, make the space changing, perpendicular to the wall body light lines extending to the top of smallpox, forming strong lighting effects, highlighting the indoor atmosphere. Leisure package box corridor wall body, champagne gold ha row mountain stage light, light and shadow of the perfect combination of shape of modern science direction, letter contact counter design concept. Restaurant concept from traditional Chinese with novel, with glass pattern represents the traditional wood carving, distributed in different location, elegant and not insolent, plus a rich focus of circular chandelier, series together, creating a oriental charm of a comfortable dining atmosphere. Box, rosewood floor and wall body, composed of Chinese style, with pattern black mirror, derived from New Chinese air mass.

Suite using both sides of the low ceilings, catch soft lighting to highlight the bed back shape, make the top of the not regular arc pattern smallpox is more interesting, not rigid, different from the traditional hotel rooms.

This is a charming The Inn Boutique, is the personality of the expression, showing the design of the not outdated. No matter how many years of washing, still fashion.

商务酒店

现代典雅

酒店入口大堂的焦点是天花八角形锥体吊灯,聚焦点指向大堂中央。沙发等候区以波浪形的地台边引领客人走到接待柜台。接待柜台后利用印花铝板面料及不规则暗灯光斜线,让平实的墙身充满艺术气质。大堂中央安置一张高贵的圆形座椅,中心位加上花艺,形成一个艺术摆设。天花的层叠造形酷似一朵朵盛放的玫瑰。西餐厅入口两侧的弧形屏风配以灰玻璃及镜钢柱,增添透视感。碍眼的建筑柱身,用古铜铝包裹成鹅蛋形,使柱身自然地引入空间当中,亦是一个正门入口指示。餐厅内的奶白色天花,设计以水珠涟漪图案为中心,下方设置自助餐台,天与地的配合,令其成为西餐厅的活动中心点。来到会议室,弧形天花中央的镜钢灯槽配合两旁花纹装饰板,就像天外有天的太空景象,亦与地台的弧形图案互相呼应,令视觉效果更有立体空间感。

在休闲中心,接待处的概念以弧线为主,使空间变化多端,墙身横纹灯光线条延伸至上方天花,形成强烈的灯光效果,突显室内气氛。

中餐厅的概念由传统中带点新颖,以图案玻璃代表传统木雕刻屏风,分布于不同位置,淡雅而不张狂,加上多个富焦点性圆形吊灯,串连在一起,营造出东方韵味的舒适用餐气氛。套房利用两旁的低天花,搭上柔和灯光框架突出床背造形,使天花的不规则弧形图案显得更有趣味,不刻板,有别于传统酒店客房。

这是个充满魅力的精品酒店,是个性的表达,展现出不过时的设计美学。无论经过多少岁月洗涤,依然时尚永恒。

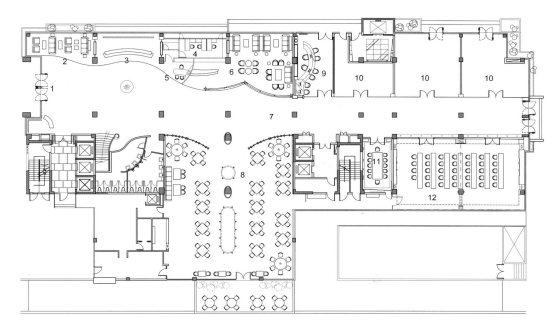

1F

1. Entrance Lobby 入口大堂
2. Seating Area 等候厅
3. Reception 接待处
4. Office 办公室
5. Lobby Manager 大堂经理
6. Coffee Shop 咖啡厅
7. Hallway 走廊
8. Restaurant 餐厅
9. Business Center 商务中心
10. Shop 店铺
11. Conference Room 会议室
12. Auditorium 大会堂

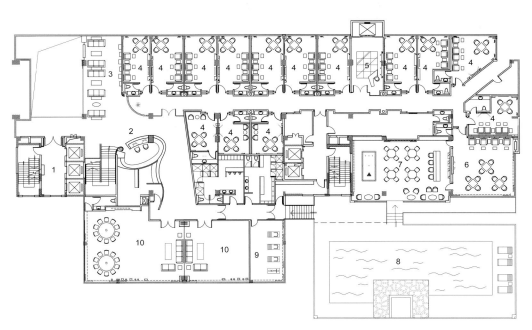

2F

1. Lift Lobby 电梯大堂
2. Counter 接待柜台
3. Seating Area 等候处
4. Mahjong & Foot Massage Room 麻雀，脚底按摩室
5. Reception 接待处
6. Grand Mahjong Room 贵宾麻雀室
7. Bar & Snooker Room 酒吧，桌球室
8. Swimming Pool 游泳池
9. GYM 健身房
10. Multi-Function Room 多功能厅

KALEIDOSCOPIC SILHOUETTE

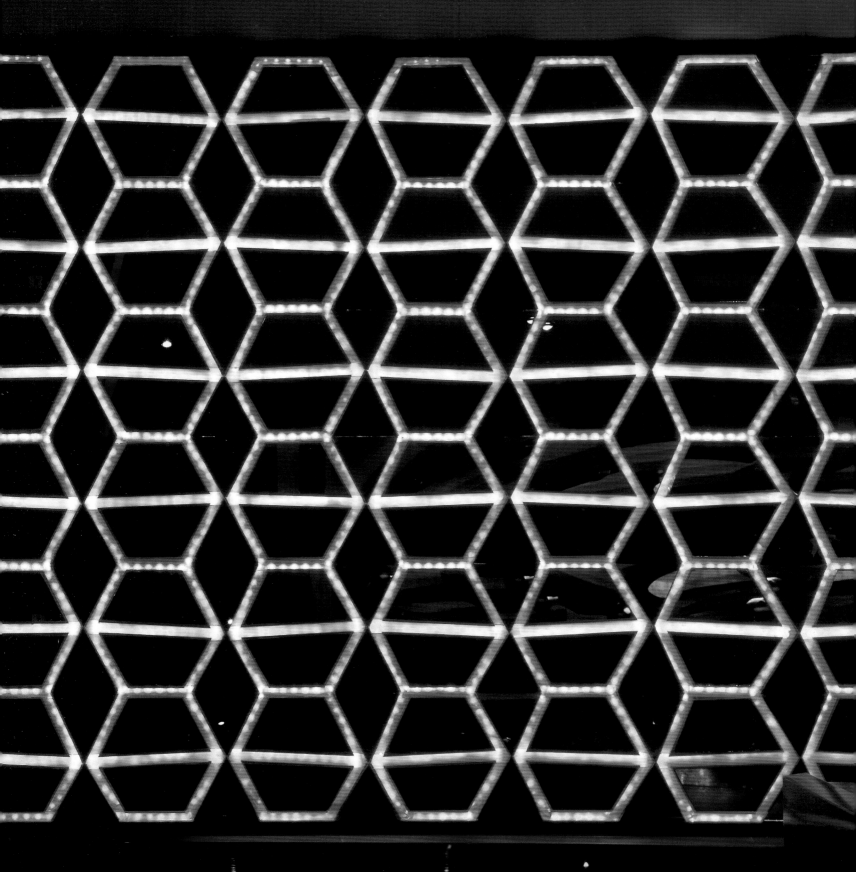

Karaoke

The design aims at creating a world of fantasy in which visitors may feel relaxed and enjoy a marvelous time of their lives. The dynamic sense was created by stunning lighting effects, complemented by other reflection elements, such as mirror polished stainless steel, glass and glistening fabrics. The signboard in a 'playing cards' pattern rimmed with LED of different colors may be noticed at a distance. From the tunnel-like entrance, through the main hall and the corridors into the VIP rooms, visitors may feel the cohesive power of strong visual effect produced by decorative LED strips lighting.

光影卡啦 OK
缤纷剪影

设计主要为了创造一个幻想的世界,让人可以从中舒缓压力,放松自己,从而享受生活的美好时光。

五光十色的奇幻世界是由多变的灯光效果,辅以其他反射元素组合而成,包括镜面抛光、不锈钢反光,玻璃折射和闪亮的面料。

吸人眼球的大招牌,采用了不同颜色的 LED 灯拼合而成,造成一副朴克牌的图案外形;从隧道般的入口,经走廊进主大厅及贵宾室,当中会感受到 LED 灯的多变迷幻的视觉效果。

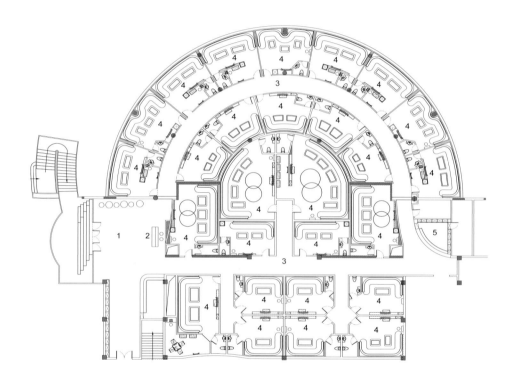

1. Entrance Lobby 入口大堂
2. Reception 接待处
3. Corridor 走廊
4. Cubicle 包厢
5. Snack Bar 小食亭

/ 306

ABSTRACT EXPRESSION

Modern Office

Abstract: To appreciate the various appearances brought by the changes and transformation in space.

Visual: White is the main colour tone and it is complemented by grey colour. The light on the wall reflects different levels of height. The set of cloud-shape ceiling light produces the infinite variations of light, and a dramatic yet balancing effect.

Curved lines: The design was based on careful and precise calculation of the curve pattern – inward and outward curved lines join the reception centre with the other four areas. There, the reception sofas and the walls blend together perfectly. The grey-colour partitions and the embedded magazine rack deliver a sense of beauty of gentle waves. The beautifully sculptured reception counter forms the focal point of the whaole area.

Zigzag lines: Beyond the double doors to the conference room, the light box in zigzag form was mounted to the false ceiling. The lighting connects the different parts of the interior bringing out an easy and interesting effect.

Interaction: In the same interior, the curve-lined lobby and the zigzag working space integrate into one whole unit fully interactive.

现代办公室

简约抽象

抽象自然： 体会空间随着改变及转换而产生更多的变化

视觉： 白色简单主色调，灰色的衬托，墙身光线高中低变化，云端的天花灯带影射出不同层次而又平行视觉的效果，永恒的变幻。

弧线： 设计以严谨的弧形比例，用内外弧线的表现方式连接招待中心的四个区域，等候沙发与墙身浑为一体。灰色玻璃的隔间及崁入式的杂志架，营造出波浪的美感，而雕塑形的接待柜台则成为整个欣赏空间的中心焦点。

折合线： 打开会议室的双掩门，折合线的天花灯箱把四平八稳的空间柔合起来，收起严肃，打开轻松又富趣味的效果。

碰合： 在同一空间范围，走道由弧线墙身带动，曲直天花墙身的工作室，融入为一整体，互动的呼应。

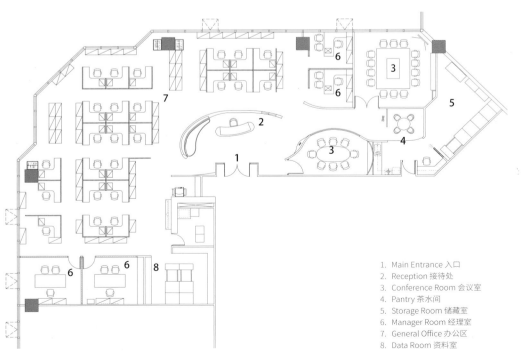

1. Main Entrance 入口
2. Reception 接待处
3. Conference Room 会议室
4. Pantry 茶水间
5. Storage Room 储藏室
6. Manager Room 经理室
7. General Office 办公区
8. Data Room 资料室

SIMPLICITY & ELEGANCE

Commercial Office

The design principle of the office was to show its vigor and the extraordinary taste of its owner. By using white colour, it creates a sense of quietness and freshness which helps bringing out the feel of vast openness of the reception hall. The wall is covered with irregular shape of wood strips of different colours.

Another focus of the project was the four Executive offices and conference room. The design took into account that an office would be a permanent or stable working environment. The design emphasizes on the use of light colours. The ceiling was left as white, and section of the walls were wrapped in a layer of pale coloured leather. Frosted and transparent glass was also incorporated into partitions. Among them, only the wooden bookshelves and office tables were left in its natural dark tone. The final result of sunlight and dark combinations creates a balanced working environment.

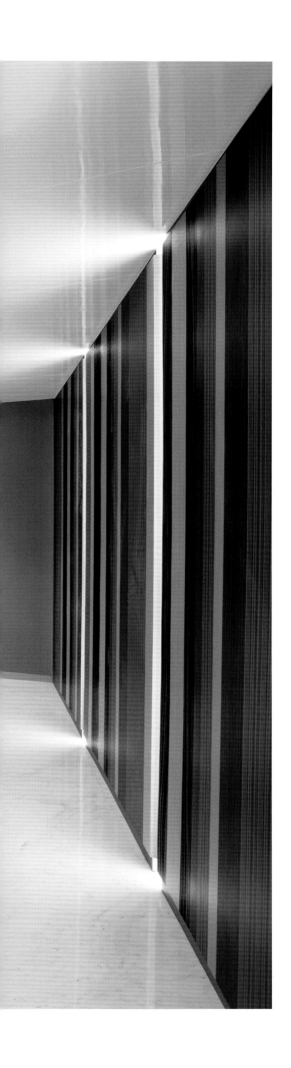

商务办公室

清雅脱俗

这所写字楼展现出了其不平凡的实力和品味，设计从这一原则出发，以白色为主色调，塑造了恬静，开阳及清新感，亦提炼出接待大堂的广阔空间。墙身的直纹间条型不规划木纹传递了动感连贯性，毫不刻板，色调变化带出趣味效果。

另一个焦点是截然不同的四间总经理室及会议室，设计偏向稳重的工作环境，天花留白，墙身扪皮板色调很浅，只有实木文件书架及工作台有啡色元素色调，部份间隔是透明感的磨纱玻璃，将空间视觉拉阔，综合以上的元素，房间塑造出和谐自然，恰到好处的舒适工作环境。

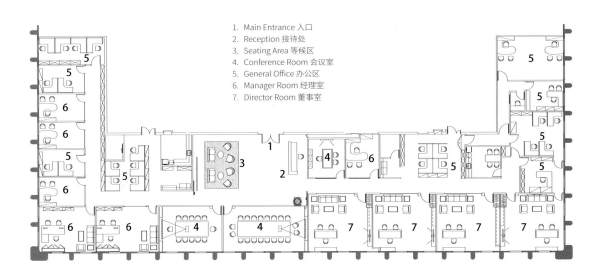

1. Main Entrance 入口
2. Reception 接待处
3. Seating Area 等候区
4. Conference Room 会议室
5. General Office 办公区
6. Manager Room 经理室
7. Director Room 董事室

MEDITATION COTTAGE

Sanji Temple

The primary design concept was tended towards the grand simplicity of the Tang Dynasty, with an aspiration to promote Chinese culture. The design also aimed at making a breakthrough in designing religious building in a more contemporary manner conforming to the natural environment. First, it is not encircled by walls so people may see it from the outside; second, it is simply furnished and decorated to give a sense of Zen; last but not least, the project has been completed economically.

三济祠

禅修道场

道场设置于后山脉脊线上,以"二进二院二门"式的设计布局。

因循地势斜度,左方长梯为"第一进",梯级旁配置级形花槽"一院",串连矮围墙,推动着长梯的牵引力互动力,加强主建筑物的方向引导。露天长梯前方设小玄关及闸门为"一门",给信众们暂停脚步,凝望庄严清静的道场。经长梯的带领,抵达主建筑露天玄关为"二进",玄关空旷平静,与天地融合,令信众进入道场前,洗涤心灵,平和心境的过渡空间。

设计规划在约4米高玄关平台前墙身,用粗犷直条形浑凝土建造,坚固有力的基础令置身上方的建筑物顿时发出磅礴的气势,与下方"二院"绿林配对,充满生机。主建筑道场中轴位入口为"二门"所在,打开双掩大门,对正就是祭坛中心,给信众参拜、修持……

整个规划分成高中低三级平台为基础,绿化的柔与主建筑、斜坡、长梯的刚互相串连,高低平衡,令人领略到地势与空间的协调,感受重重无尽的"禅"的意味。

/ 331

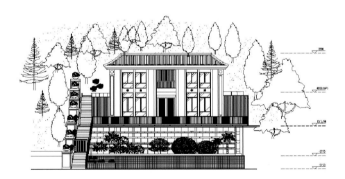

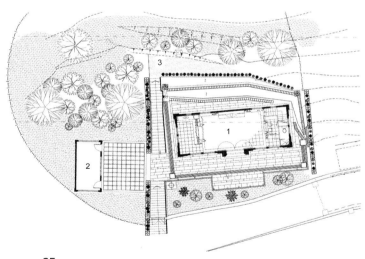

GF

1. Main Building 主建筑物
2. Garden Shed 小屋
3. Backyard 庭院

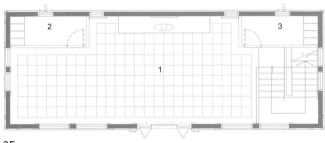

2F

1. Meditation Hall 静心堂
2. Male Dressing Room 男更衣室
3. Female Dressing Room 女更衣室

主创设计师 Main Design Team

王启贤 Clyde Wong
杨秉昆 Winston Yeung
曾皓柃 Ling Tsang
许佰顺 Parson Hui

Established in 1993, WIA Design Consultants has gained invaluable experience in the industry over the years. With more than 30 years of experience, our Design Director, Mr. Clyde Wong together with his team members is able to deliver innovative and distinguished designs for our clients in a professional manner.
FROM IDEA TO REALITY

王启贤设计事务所
WIA DESIGN CONSULTANTS

香港湾仔告士打道 128 号 12 楼 F 室
Unit F, 12/F Neich Taver, 128 Gloucester Road, Wan Chai, Hong Kong
TEL: (0852) 2549-0228
EMAIL: wiacph@netvigator.com

香港 · 广州 · 深圳 · 西安

www.wiadesign.com

创福美图 名师专辑出版策划，请洽翟生 138-0226-6970